Advances in Intelligent Systems and Computing

Volume 304

Series editor

Janusz Kacprzyk, Polish Academy of Sciences, Warsaw, Poland
e-mail: kacprzyk@ibspan.waw.pl

About this Series

The series "Advances in Intelligent Systems and Computing" contains publications on theory, applications, and design methods of Intelligent Systems and Intelligent Computing. Virtually all disciplines such as engineering, natural sciences, computer and information science, ICT, economics, business, e-commerce, environment, healthcare, life science are covered. The list of topics spans all the areas of modern intelligent systems and computing.

The publications within "Advances in Intelligent Systems and Computing" are primarily textbooks and proceedings of important conferences, symposia and congresses. They cover significant recent developments in the field, both of a foundational and applicable character. An important characteristic feature of the series is the short publication time and world-wide distribution. This permits a rapid and broad dissemination of research results.

More information about this series at http://www.springer.com/series/11156

Rituparna Chaki · Khalid Saeed
Sankhayan Choudhury · Nabendu Chaki
Editors

Applied Computation and Security Systems

Volume One

Springer

Editors
Rituparna Chaki
A.K. Choudhury School of Information
  Technology
University of Calcutta
Kolkata, West Bengal
India

Khalid Saeed
Faculty of Physics and Applied Computer
  Sciences
AGH University of Science and Technology
Cracow
Poland

Sankhayan Choudhury
Nabendu Chaki
Department of Computer Science
  and Engineering
University of Calcutta
Kolkata, West Bengal
India

ISSN 2194-5357
ISBN 978-81-322-1984-2
DOI 10.1007/978-81-322-1985-9

ISSN 2194-5365 (electronic)
ISBN 978-81-322-1985-9 (eBook)

Library of Congress Control Number: 2014947644

Springer New Delhi Heidelberg New York Dordrecht London

Printed on acid-free paper

Springer is part of Springer Science+Business Media (www.springer.com)

Preface

The First International Doctoral Symposium on Applied Computation and Security Systems (ACSS 2014) took place during Apr 18–20, 2014 in Kolkata, India. This symposium is aimed to facilitate the Ph.D. students to present and discuss their research work leading towards high-quality dissertation. This symposium will provide a friendly and supportive environment for doctoral students to present and discuss their work both with their peers and with a panel of distinguished experts. ACSS Doctoral Symposium allowed researchers working in different fields of computer science such as Image processing, Remote Healthcare, Biometrics, Pattern Recognition, Embedded Systems, Data Mining, Software Engineering, Networking, and Network Security. The symposium evolved as a joint venture between two collaborative universities: the University of Calcutta, India, and the AGH University of Science and Technology, Poland.

The program committee members of ACSS 2014 were instrumental in disseminating the objectives of the symposium among the scholars and faculty members in a very short time. This resulted in a large number of submissions from Ph.D. scholars from India and abroad. These papers underwent a minute and detailed blind-review process with voluntary participation of the committee members and external expert reviewers. The metrics for reviewing the papers had been mainly the novelty of the contributions, technical content, organization, and clarity in presentation. The entire process of initial paper submission, review, and acceptance were done electronically. The hard work done by the Organizing and Technical Program Committees led to a superb technical program for the symposium. The ACSS 2014 resulted in high-impact and highly interactive presentations by the doctoral students.

The Technical Program Committee for the symposium has selected only 25 papers for publication out of a total 70 submissions. Session chairs were entrusted with the responsibility of submitting feedbacks for improvements of the papers presented. The symposium proceeding has been organized as a collection of papers, which were presented and then modified as per reviewer's and session chair's comments. This has helped the scholars to further improve their contributions.

We would like to take this opportunity to thank all the members of the Technical Program Committee and the external reviewers for their excellent and time-bound review works. We especially thank Prof. Indranil Sengupta of IIT, Kharagpur for his suggestions towards designing the Technical Program for ACSS-2014. We thank all our sponsors who have come forward towards organization of this symposium. These include Tata Consultancy Services (TCS), Springer India, ACM India, M/s Business Brio, M/s Enixs. We appreciate the initiative and support from Mr. Aninda Bose and Ms. Kamiya Khatter his colleagues in Springer for their strong support towards publishing this post-symposium book in the series "Advances in Intelligent Systems and Computing." Last, but not the least, we thank all the authors without whom the symposium would not have reached up to this standard.

On behalf of the editorial team of ACSS 2014, we sincerely hope that the different chapters of this book will be beneficial to all its readers and motivate them towards further research.

Rituparna Chaki
Khalid Saeed
Sankhayan Choudhury
Nabendu Chaki

Contents

About the Editors

Rituparna Chaki is an Associate Professor in the A.K. Choudhury School of Information Technology, University of Calcutta, India since June 2013. She joined the academia as faculty member in the West Bengal University of Technology in 2005. Before that she has served under Government of India in maintaining industrial production database. Rituparna has done her Ph.D. from Jadavpur University in 2002. She has been associated in organizing many conferences in India and abroad as Program Chair, OC Chair, or as member of Technical Program Committee. She has published more than 60 research papers in reputed journals and peer-reviewed conference proceedings. Her research interest is primarily in Ad-hoc networking and its security. She is a professional member of IEEE and ACM.

Khalid Saeed received the B.Sc. Degree in Electrical and Electronics Engineering from Baghdad University in 1976, the M.Sc. and Ph.D. Degrees from Wroclaw University of Technology, in Poland in 1978 and 1981, respectively. He received his D.Sc. Degree (Habilitation) in Computer Science from Polish Academy of Sciences in Warsaw in 2007. He is a Professor of Computer Science with AGH University of Science and Technology in Poland. He has published more than 200 publications—edited 23 books, Journals and Conference Proceedings, 8 text and reference books. He supervised more than 110 M.Sc. and 12 Ph.D. theses. His areas of interest are Biometrics, Image Analysis, and Processing and Computer Information Systems. He gave 39 invited lectures and keynotes in different universities in Europe, China, India, South Korea, and Japan. The talks were on Biometric Image Processing and Analysis. He received about 18 academic awards. Khalid Saeed is a member of more than 15 editorial boards of international journals and conferences. He is an IEEE Senior Member and has been selected as IEEE Distinguished Speaker for 2011–2016. Khalid Saeed is the Editor-in-Chief of International Journal of Biometrics with Inderscience Publishers.

Sankhayan Choudhury is Associate Professor in the Department Computer Science and Engineering, University of Calcutta, India. Currently, he is head of this department. Moreover, he is Co-ordinator of TEQIP-II, University of Calcutta. Dr. Choudhury has obtained his B.Sc. (Hons.) in Mathematics under University of Calcutta. Thereafter he has obtained B.Tech. and M.Tech in Computer Science and Engineering from University of Calcutta. He has completed Ph.D. from Jadavpur University, India in 2006. His research interests include Mobile Computing, Networking, Sensor Networking, Cloud Computing, etc. Besides authoring a book, Dr. Choudhury has published close to 50 peer-reviewed papers in international journals and conference proceedings. He has also served in the Program Committees of several international conferences and has also chaired the Program and Organizing Committees of a few. Dr. Choudhury is a professional member of ACM and an executive committee member for the local ACM professional chapter in Kolkata, India.

Nabendu Chaki is a Senior Member of IEEE and an Associate Professor in the Department Computer Science and Engineering, University of Calcutta, India. Besides editing several volumes in Springer in LNCS and other series, Nabendu has authored three textbooks with reputed publishers like Taylor and Francis (CRC Press), Pearson Education, etc. Dr. Chaki has published more than 120 refereed research papers in Journals and International conferences. His areas of research interests include image processing, distributed systems, and network security. Dr. Chaki has also served as a Research Assistant Professor in the Ph.D. program in Software Engineering in U.S. Naval Postgraduate School, Monterey, CA. He is a visiting faculty member for many universities including the University of Ca'Foscari, Venice, Italy. Dr. Chaki has contributed in SWEBOK v3 of the IEEE Computer Society as a Knowledge Area Editor for Mathematical Foundations. Besides being in the editorial board of Springer and many international journals, he has also served in the committees of more than 50 international conferences. He is the founding Chapter Chair for ACM Professional Chapter in Kolkata, India since January 2014.

Part I
Pattern Recognition

An Algorithm for Extracting Feature from Human Lips

Joanna Kosior, Khalid Saeed and Mateusz Buczkowski

Abstract This paper presents a new method for extracting features from human lips. Correct pointers extraction has a significant meaning for the whole process of identification, recognition expressions and detection of people features. The introduced algorithm concerns about finding four points around the mouth: two for corners, one situated in center on the border of upper lips and the last on the border of lower lips, which next are used for creating feature vectors

Keywords Biometrics · Feature extraction · Feature vectors · Mouth biometric features · Human feature extraction

1 Introduction and State of the Art

Modern technology allows us to construct the device of unimaginable power. Because of this, it becomes possible to implement complex algorithms previously requiring too much computing time. Programs of this type are methods that recognize images and extract from them interesting to us items. Now we can take a picture of any object, string or a person and the right program is able to interpret

J. Kosior (✉) · K. Saeed · M. Buczkowski
Faculty of Physics and Applied Computer Science, AGH University of Science and Technology, Krakow, Poland
e-mail: joanna.kosior0@gmail.com

K. Saeed
e-mail: saeed@agh.edu.pl
URL: http://home.agh.edu.pl/~saeed/

M. Buczkowski
e-mail: mateusz.buczkowski1@gmail.com

K. Saeed
Faculty of Computer Science, Bialystok University of Technology, Bialystok, Poland

© Springer India 2015
R. Chaki et al. (eds.), *Applied Computation and Security Systems*, Advances in Intelligent Systems and Computing 304, DOI 10.1007/978-81-322-1985-9_1

for us the image. Finds the faces in the picture, reads the bar code or even search the web string of characters that made it onto the photographed device. The spectrum of applications is wide, stretches from the needs of military to social networking sites. Among these, the last is particularly important to recognize and identify people and their emotions. Human face is very expressive, expresses many different emotions. Mimicry provides a lot of information about the person. For this purpose, we use algorithms for classifying facial expressions as appropriate emotions. That is why is needed the widest possible development of this field.

One of the most amazing parts of human abilities is recognizing expressions from human faces [1, 2]. Human brain can assess data from images acquired by eyes. There are several approaches to detect face and hence eyes, nose and lips [3, 4, 5]. The following work leads to human lips location. The presented by authors method creates and uses a vertical histogram to find the mouth area in which are searched the corners and the lower and upper lip points. With these, points are created feature vectors used in the classification process of facial expressions. To get the correct position of the mouth, the histogram must first find the area of the face and bring it to a binarized picture containing as little as possible details of the face surrounding and as many as possible details from the areas of the mouth, nose and the eyes.

The method described in [6] based on Gabor filter is one of the most effective methods for extracting features. It is usually used for facial expression recognition and gives very good results. Due to the huge amount of information production for its implementation, we need additional data selection AdaBoost and/or SVM algorithm. Due to the lower amount of data and a small Gabor filter effectiveness in the area of the lips alone, the authors in [6] proposed a method PHOG. This method, in opposed to Gabor filter, does not recognize all the emotions, realizes only detection of the mouth and the smile recognition. It is much faster and reaches a similar effectiveness. In the first step, Canny filter is used to detect lips edges and hence the contour. Then, the image is divided into cells. In each successive step, the increasing number of parts in the horizontal and the vertical directions is 2^x cells. The method of subdivided image parts is shown in Fig. 1.

The method counts the direction of the gradient in the subsequent parts of the picture, and on this basis, HOG—histogram of orientation gradients is formed (Fig. 2).

In [7] has been described the recognition of emotions per the help of both a frontal view of the face and its profile. At the beginning, segmentation is performed on the basis of the color image, finding the area of the skin as the largest connected with each other set of pixels: $H = [\max(-0.7, Hs - 0.35), \min(Hs + 0.35, 1.05)]$, $S = [0, 0.7]$, $V = [0.1, 0.9]$. Where H, S and V are the variables of the model HSV, and Hs is the average hue H. After the operation of erosion, watershed segmentation is used to find the area of the face. Searching this area, the authors in [7] relied on the use of watershed algorithm to save time implementation of the program. Example results of successive steps are provided in Fig. 3.

In order to find the best fitting points that characterize a given feature, several approaches were used to check which works best for a particular feature.

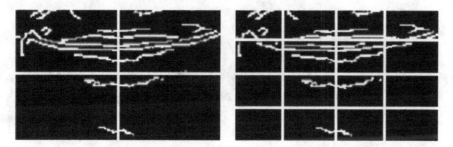

Fig. 1 Dividing the image into cells [6]

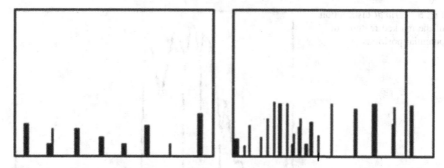

Fig. 2 On the *left* is the HOG for partitioning the image into 4 cells and on the *right* 16 cells [6]

Fig. 3 The steps of face area search [7]

Two detectors were selected: curve fitting—parabolas to fit the contour of lips and dimensional model of the mouth. For further analysis, these four points will characterize the expression of the mouth (Fig. 4).

The method of extracting the face in [8] focuses on the analysis of skin color in the HSV color space, thanks to this one can find the colors which correspond to fields of color face. The next step is to fit an ellipse to the contours of the face based on his skin regions received. It was observed that facial features have a lower than the areas of the skin brightness. On this basis was created the vertical distribution of gray levels in a row. The resulting shape is rotated to normalize the ellipse face. Minima and maxima correspond to a specific facial features (Fig. 5).

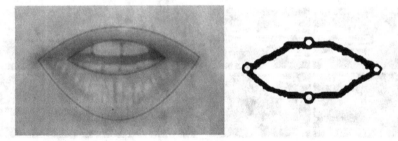

Fig. 4 A typical result of matching the shape of the mouth with extracted points [7]

Fig. 5 *Vertical* distribution of the marked features of potential positions [8]

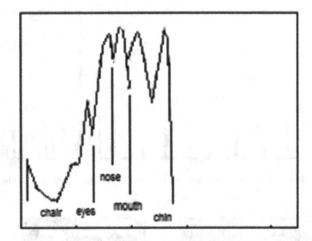

The next step is to create a horizontal distribution for each minimum and maximum vertical distribution. In this way, we obtain the position of the data features on the face. In (Fig. 6), minima can be seen from most likely from the eyes.

At matched pairs of horizontal distribution of minima which correspond best to the requirements of the features used are the base of the used fuzzy sets-based theory. For each feature, a function is defined. On the basis of experimentally determined parameters, it is decided which positions are acceptable and will be assigned to features. Then the points are grouped and the center of each of these groups is designated. The effects are shown in Fig. 7.

2 Algorithm

2.1 Image Database

The database photographs used in the tests is FEI Face Database [9]. It contains 200 color photographs of human faces. All images are of size 360 × 260 pixels.

Fig. 6 An example of *horizontal* distribution for minimum *vertical* distribution [8]

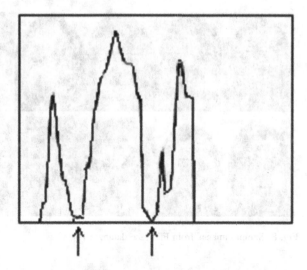

Fig. 7 Detection of nose, mouth and chin positions [8]

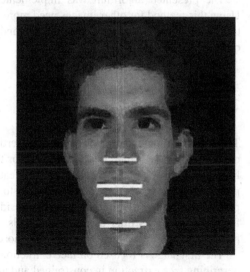

Every person has taken two pictures: one with a smile on the face and the other with a neutral expression. The position of the face at each photograph is normalized—every face has the same size in the picture. The eyes are located on the same or in a few cases, on a similar height. The same applies to the nose and mouth. In the horizontal direction is also centered face, the area of mouth is contained in the range [0.3:0.7] image width. Background is always a light gray color. Generally, the faces are completely exposed, but some of men's have beard and mustache which impact upon properly finding mouth area. The presence of other objects for example glasses does not have any affection for final result. Sample images from the base are presented below in Fig. 8.

Fig. 8 Sample images from FEI face database [9]

The presented algorithm was implemented exactly for FEI Face Database. Using this method for other images need more development and it is not a topic for this article. Images have to be normalized and have uniform lighting and uniform quality.

2.2 Face Extraction

The presented method creates and uses a vertical histogram to locate the position of the lips. At the found area is searched for the corners and the lower and upper lip. From the marked points can be created feature vectors used in the classification process facial expressions. To obtain a histogram giving the correct location of paragraphs need to first find the area of considered face and bring it to a binarized form containing as little as possible details of the background and as much as possible details of the area of the mouth, nose and eyes (Fig. 9).

The first stage is focused on face extraction. There are known many algorithms describing face extraction in constrained and also in unconstrained conditions [10]. Below, we present the method operating in constrained conditions. In the initial image preparation, histogram normalization is applied to the image to obtain a better contrast, which should contribute to increase the effectiveness of finding the area of the face. This procedure extends a histogram. Figures 10 and 11 show the results of this stage.

The next step is to obtain an image of the subtracted channels R–G through actions on the RGB channels of image. Channels R—red and G—green contain much information about the difference of color of the face and the environment. For each pixel in the image is taken red and green channels, in order to subtract them from each other. If the result is less than zero, a new pixel color values, zero

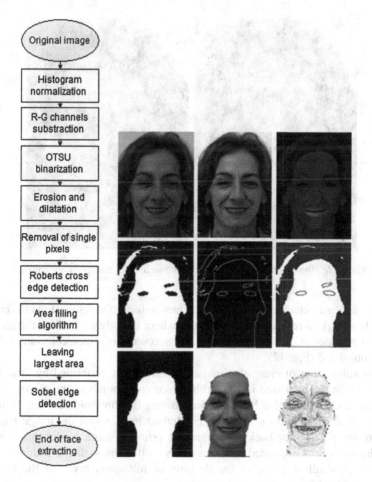

Fig. 9 Diagram of operation extracting face on the *left*, on the *right* next steps to extracted face

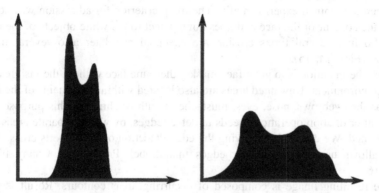

Fig. 10 On the *left* the original histogram, on the *right* histogram after normalization

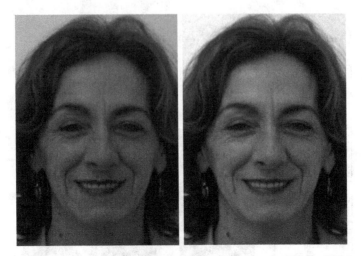

Fig. 11 On the *left* the original image, on the *right* image after histogram normalization

is inserted into all channels, otherwise the new value is the result of the difference channels RG. As a result, we get a picture where the brightest colors are marked with the prevalence of skin areas with a low content of green color and a large saturation of red (Fig. 12).

After subtraction of channels, image thresholding is performed by Otsu algorithm [11]. In the binarized image, white color corresponds to the places of skin occurrence. Otsu method is based on determining the threshold resulting from the analysis of the histogram of the image. Method implies that the image has two states of the pixels—the background and the primary image. Under this assumption, the histogram is divided into two states so that the variance within each of them is the smallest. Place of the division of histogram becomes the adopted threshold (Fig. 13).

To smooth edges of the face and reduce the impact of larger artifacts, the operations of erosion and dilation are used. The order to perform these operations has been determined experimentally. The main criterion for admission was not to break the contour of the face and does not connect to the white objects of the area face. So first several times erosion are executed and later also several times dilation (Figs. 14, 15).

Now the intention is to get a face mask—the same face without the visible area of the environment. Unwanted areas are also located within the remains of the face such as the eyebrows, nose, eyes, mustache, mouth or chin. For this purpose, the image after dilation operations needs to detect edges, by which separate areas can be obtained. We do this by applying the edge detection of the Roberts cross. This filter allows us to get narrower edges than Sobel, Prewitt and Canny filters (Fig. 16).

The resulting image is composed of occurring areas contours. Result is presented in Fig. 17.

Fig. 12 Image after subtraction of R and G channels

Fig. 13 The image obtained by Otsu binarization

Now, we make a negative image and with the area filling algorithm (called flood fill or seed fill [12]) we fill in the red first the encountered area from the upper left corner. This is the background area. This has been implemented on the basis of the queue version floodfill area filling algorithm. The method requires three parameters: the initial position, replaced color and the new color. It is the four-directional version of floodfill algorithm. This variation of the algorithm does not "overflow" by diagonal lines at an angle of 45°. In Fig. 18, this is marked with black edges. Each of the areas designated by them is filled with a different color.

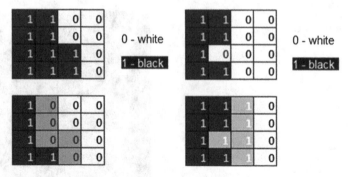

Fig. 14 Idea of erosion and dilation operations

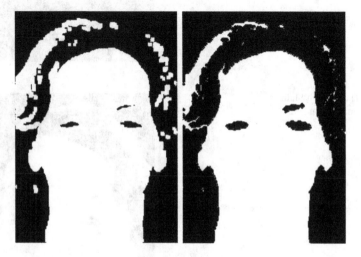

Fig. 15 Face area after erosion on the *left*, and on the *right* face area after dilation

Thanks to this, we can specify the size of the area. We leave the area with the largest field and darkening the rest into black Fig. 19. The last operation needed to create a mask of the face is dilation of the image obtained. After this procedure, facial area will be narrower and when the face will be isolating the hair, the edges of the head and the other image objects that do not belong directly to the face will not be included.

At this point, we are only interested in the location of individual facial features discarding the colors. To extract the mouth, eyes and nose, we need to apply the filter edge. In this case, the best results are achieved by Sobel edge filter. Thanks to this all the edges are found on the image. Apart from main outlines will be visible wrinkles while the dominant edge density areas will be highlighted by thicker lines of white color (Fig. 20).

Fig. 16 Presented comparison between Roberts cross (*left*) and Sobel edges filter (*right*)

Fig. 17 Detected edge—
image with contours of face
and other remaining objects

2.3 Mouth Extraction

The next step is finding and extracting the mouth area. Based on binarized image is calculated vertical intensity histogram of black pixels in each row, counting the number of black pixels, saving the greatest value and normalizing all the remaining ones with respect to it (by dividing by its value). Using this data for each row, horizontal lines are drawn with its length being the intensity value of black pixels in that row divided by the largest intensity value taken from all rows.

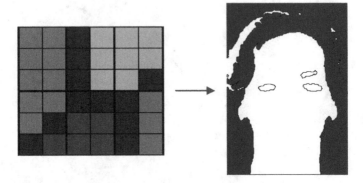

Fig. 18 Illustration of the floodfill function operation [12]

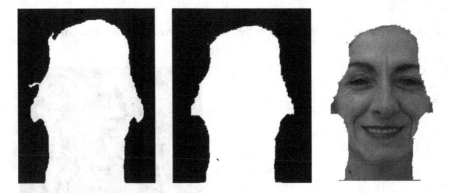

Fig. 19 Created mask, mask after dilation, face cut by mask

Fig. 20 Image of face after
Sobel filter, all feature are
visible and expose

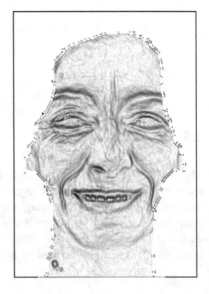

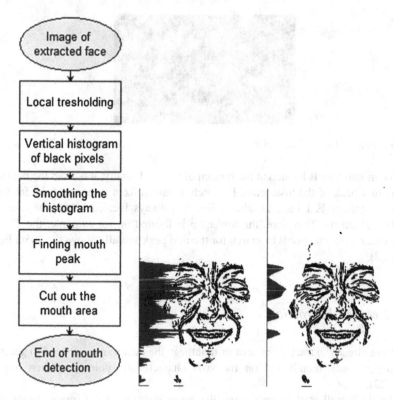

Fig. 21 On the *left jagged shape* of histogram, on the *right* improved shape

The resulting histogram is strongly jagged in shape, and it is difficult to read the necessary information from it (Fig. 21).

To smooth the histogram and hence to keep only the large peaks, the length of each line is calculated as the weighted average of the nearest lines obtained from Eq. (1).

$$F(j) = [3F(j-4) + 3F(j-3) + 2F(j-2)$$
$$+ 3F(j-1) + 2F(j) + 3F(j+1) + 2F(j+2) \qquad (1)$$
$$+ 3F(j+3) + 3F(j+4)]/25$$

where

j Line number
$F(j)$ Number of black pixels in row j

The division by 25 is for normalization.

In addition, in order to remove noise in the form of very low peaks, in each line is removed as 0.45 times the value of the maximum counts for all rows.

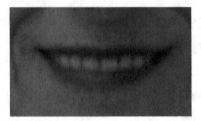

Fig. 22 Extracted area of the mouth

The mouth area is located at the bottom of image. Usually, it is from the bottom of the first peak of the histogram. For such a standardized position of the face as the base gallery FEI Face Database, lips are always located below the line of height 305 pixels. Therefore, the lower lip is located in, for example, the 290th line. This means one needs to search for the first peak visually above the 305th line (Fig. 22).

2.4 Feature Extraction

We have already extracted the area of mouth. In the next move, we need to get the contour of only mouth and on its basis characteristic points are then found (Fig. 23).

The work will start by extracting the green channel of the image. From the experimental results, the mouth image is most diverse precisely in the background of skin when taken from this channel. The middle image (green channel) in Fig. 24 is illustrating this fact.

The resulting image is changed to grayscale and then binarized with mixed thresholding with the following parameters: the width of the environment—15 pixels, 35 percent degree of deviated from the global average and a threshold of 60 on the [0:255] scale. All parameters are selected experimentally—they have given the best results for this database (Fig. 25).

Then to only get the contour of the lips on the negative image each white area is given a different color $(R + i, 0, 0)$, where i is the number (position) of the next area. Filling and counting fields follows by floodfill algorithm (Fig. 26).

In the next step, the rest of smaller areas is obscured. Background color is changed into white and the area of the lips is then colored black (Fig. 27).

This way the mouth outline is obtained, on which the position of the lips corners and the farthest points on the lower and upper lips are easily found. The best method of finding the corners of the mouth turned out to be the extraction of the G channel image of the mouth image after converting into grayscale. The darkest points in the image are the corners of the mouth, their brightness represents 13 % of the maximum brightness, that is in RGB 255 G, which gives 33 G corner color. These areas are found—they are the farthest to the left and to the right

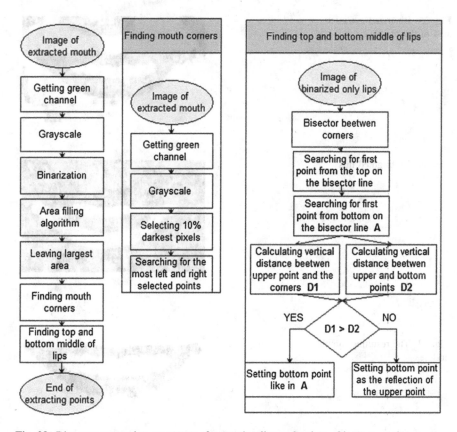

Fig. 23 Diagrams presenting next steps of extracting lips and points of interest on them

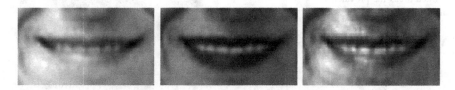

Fig. 24 Extracted area of the mouth, from the *left*: *red*, *green* and *blue* channels presented in *gray* scale

points. They are marked as small light areas on Fig. 28, which shows the lips area in the mouth separated part of the face image.

With the coordinates of the corners, we find the stretch between them. On the bisector of the lips image outline the first black point from the top will indicate the searched upper farthest point. The algorithm now splits into two cases aiming at increasing the effectiveness of selecting the most extreme points of the upper and lower lips. Then:

Fig. 25 Binarized extracted
area of the mouth

Fig. 26 Highlighted lips

Fig. 27 The resulting mouth
outline

Fig. 28 The mouth area with
marked areas the prevalence
of potential corners

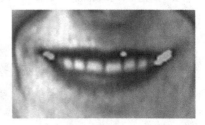

(a) we act for the lower lip the same as with the upper lip—searching for the
    bisector of the outermost corners of the bottom and selecting it as the lowest.
(b) we calculate the vertical distance point found on the upper lip to the average
    right plus left corners. New point on the lower lip is the reflection of the upper
    point.

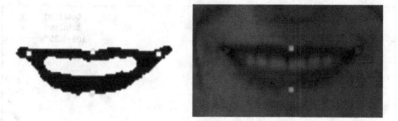

Fig. 29 Marked characteristic points of mouth

For both cases is calculated the distance of the found points on the lips vertically. The solution with the greater distance is considered (Fig. 29).

2.5 Classification

In an exemplary classification process of facial expressions, we consider two
angles determined by the position of the corners and points on the lower and upper
lip. The angles were calculated using the scalar product on the basis of previously
found points. Interesting letter angles indicated in Fig. 30.

Can be noted that for smiles scattering angles between the upper and lower lip
is significantly larger than for other expressions. Therefore, classification of facial
expressions based on finding the most optimal threshold for the module of the
difference between the calculated angles of each lips. Figure 31 shows the difference module angles.

We experimentally set the optimal threshold value 8 on the basis of achieved
73 % compatibility classification with images base (Fig. 32).

In Fig. 33, we show examples mouth well recognized as smiling and Fig. 34
example wrong diagnosis by the algorithm.

After deducting the lips of the found bad feature vectors (about 8 %) effectiveness of the algorithm of recognizing smile and non-smile faces is 84 %.

Fig. 30 Marked angles of
interest

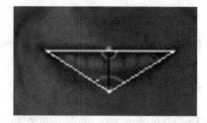

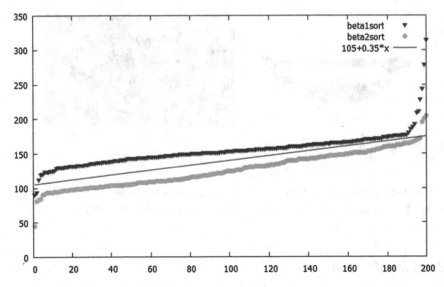

Fig. 31 Sorted collection of angles and linear function between axes

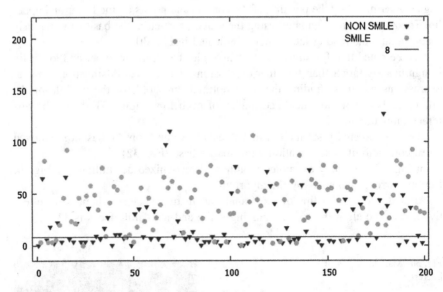

Fig. 32 Sorted collection of angles and linear function between axes

3 Experiments and Results

The algorithm was implemented in JAVA and tested on 200 images from the database. Typical results for matching characteristic points of lips are shown in Fig. 35. After finding the points by the program, verification of proper fitting is

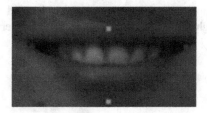

Fig. 33 Expression well recognized as smiling

Fig. 34 Expression wrong recognized as smiling

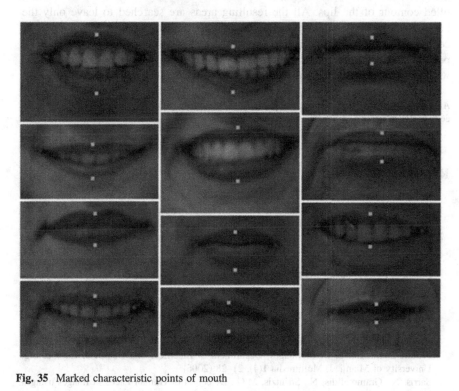

Fig. 35 Marked characteristic points of mouth

performed visually. Implemented program correctly located lips on the image in 98 % of cases, although the correct localization of characteristic points is at the level of 92 %.

4 Conclusions

The work was mainly based on the operations performed to find the face in the image. The presented algorithm starts by locating the area of the skin method based on the difference channels R–G. Then it proceeds with image binarization and edge smoothing by erosion and dilation operations. The algorithm then searches for the greatest originated areas from the face. This brings us to the point where we use edge detection algorithm and obtain the edge image of the face. On this basis is created the vertical distribution of black pixels counting. By using the weighted average, the distribution function is smoothed. Then the algorithm finds and cuts the area containing only the mouth. On the resulting distribution, the first peak from the bottom is the approximate position of the center of the lips. Using the green channel of the image, the mouth area is binarized in order to obtain a filled contour of the lips. All the resulting areas are searched to leave only the largest one—it is the area of the lips. On it are searched corners and points at the center of both lips. Two methods were worked out for locating these specific points. The algorithm counts the distance between points and accordingly decides which of them satisfy the conditions for the four characteristic points of the lips.

Acknowledgment The research was partially supported by grant No. WFiIS 11.11.220.01/sa-eed, AGH University of Science and Technology in Cracow.

References

1. Li, Y., Wang, S., Zhao, Y., Ji, Q.: Simultaneous facial feature tracking and facial expression recognition. IEEE Trans. Image Process **22**(7), 2559–2573 (2013)
2. Hirata, W., Tan, J.K., Kim, H., Ishikawa, S.: Recognizing Facial Expression for Man-machine Interaction, Department of Mechanical and Control Engineering Kyushu Institute of Technology, Japan. In: ICROS-SICE International Joint Conference, Fukuoka International Congress Center, Japan, 18–21 Aug 2009
3. Saeed, K.: Minimal-eigenvalue-based face feature descriptor. In: Dramiński, M., Grzegorzewski, P., Trojanowski, K., Zadrożny, S. (eds.) Issues in Intelligent Systems Models and Techniques. Institute of System Research, Polish Academy of Sciences, EXIT, Warsaw, pp. 185–196 (2005)
4. Mahoor, M.H., Abdel-Mottaleb, M., Ansari, A.N.: Improved Active Shape Model for Facial Feature Extraction in Color Images, Department of Electrical and Computer Engineering, University of Miami. J. Multimedia **1**(4), 21–28 (2006)
5. Sarris, N., Grammalidis, N., Strintzis, M.G.: Detection of Human Faces in Images using a Novel Neural Network Technique, Information Processing Laboratory, University of Thessaloniki, 2007. Available at http://citeseerx.ist.psu.edu/

6. Bai, Y., Guo, L., Jin, L., Huang, Q.: A novel feature extraction method using pyramid histogram of orientation gradients for smile recognition. In: 16th IEEE International Conference on Image Processing (ICIP), School of Electronic and Information, South China University of Technology, pp. 3305–3308 (2009)
7. Pantic, M., Rothkrantz, L.J.M.: Facial action recognition for facial expression analysis from static face images. IEEE Trans. Syst. Man Cybern. Part B Cybern. 34(3), 1449–1461 (2004)
8. Sobottka, K., Pitas, I.: A novel method for automatic face segmentation, facial feature extraction and tracking. Department of Informatics University of Thessaloniki, Greece, Signal Proc. Image Commun. 12(3), 263–281 (1998)
9. FEI Face Database. Artificial Intelligence Laboratory of FEI in Sao Bernardo do Campo, Sao Paulo, Brazil, 2006. http://fei.edu.br/~cet/facedatabase.html. Accessed 5 Jan 2014
10. Kocjan, P., Saeed, K.: Face Recognition in Unconstrained Environment. In: Biometrics and Kansei Engineering. Springer Science and Business Media, NY (2012)
11. Otsu method of Binarization and Thresholding. http://www.sas.bg/code-snippets/image-binarization-the-otsu-method.html. Accessed 26 Jan 2014
12. Flood Fill or Flood Seed. http://en.wikipedia.org/wiki/Flood_fill. Accessed 26 Jan 2014

Feature Selection using Particle Swarm Optimization for Thermal Face Recognition

Ayan Seal, Suranjan Ganguly, Debotosh Bhattacharjee, Mita Nasipuri and Consuelo Gonzalo-Martin

Abstract This paper presents an algorithm for feature selection based on particle swarm optimization (PSO) for thermal face recognition. The total algorithm goes through many steps. In the very first step, thermal human face image is preprocessed and cropping of the facial region from the entire image is done. In the next step, scale invariant feature transform (SIFT) is used to extract the features from the cropped face region. The features obtained by SIFT are invariant to object rotation and scale. But some irrelevant and noisy features could be produced with the actual features. Unwanted features have to be removed. In other words, optimum features have to be selected for better recognition accuracy. Since PSO is an optimization method, which works with the principle of local as well as global searches for finding optimum set of features. Here, this process has been implemented to select a subset of features that effectively represents original feature extracted for better classification convergence. Finally, minimum distance classifier is used to find the class label of each testing images. Minimum distance classifier acts as an objective function for PSO. In this work, all the experiments

A. Seal (✉) · S. Ganguly · D. Bhattacharjee · M. Nasipuri
Department of Computer Science and Engineering, Jadavpur University, Kolkata, India
e-mail: ayan.seal@gmail.com

S. Ganguly
e-mail: suranjanganguly@gmail.com

D. Bhattacharjee
e-mail: debotoshb@hotmail.com

M. Nasipuri
e-mail: mitanasipuri@yahoo.com

A. Seal · C. Gonzalo-Martin
Center for Biomedical Technology, Universidad Politecnica de Madrid, Madrid, Spain
e-mail: consuelo.gonzalo@upm.es

© Springer India 2015
R. Chaki et al. (eds.), *Applied Computation and Security Systems*, Advances in Intelligent Systems and Computing 304, DOI 10.1007/978-81-322-1985-9_2

have been performed on UGC-JU thermal face database. The maximum success rate of 98.61 % recognition has been achieved using SIFT and PSO for frontal face images and 90.28 % for all images.

Keywords Face recognition · Infrared face images · Scale invariant feature transform · Particle swarm optimization

1 Introduction

Among all biometric identification systems, face recognition is one of the most suitable methods due to its non-intrusive nature. Face recognition has appeared as one of the most exciting research problem due to its numerous practical applications. It is extensively used in the authentication, surveillance, security, and human computer interaction purpose. But the existing face recognition systems have several limitations. The performance of face recognition is outstanding in case of controlled conditions. The performances degrade significantly in an uncontrolled environment. The main reason is that the most of the existing face recognition methods are based on visual images. The quality of the visual images changes with lighting condition, as a result, the performances degrade. That means the performance of the face recognition system, based on visual images, captured in daylight situation is not same with the performance based on nightlight vision. This problem is known as illumination problem. This problem can be solved by the uses of thermal face images instated of visual images. A thermal face image is captured by the thermal infrared camera, and a thermal infrared camera is illumination independent. It can capture nearly same images in all the environments even in a dark situation because a thermal infrared camera concentrates on emitted energy from the object surface. It does not consider the environment temperature. Another significant advantage of face recognition based on thermal face images is that the tasks of face detection, location, and segmentation are relatively easier and more reliable than their visual counterpart [1].

Some general face recognition methods include eigenfaces [2, 3] and Fisher's discriminant analysis [4], which are sensitive to illumination and different facial expressions. These methods are also used for dimensionality reduction, when the dimensionality is curse for some applications. Sometimes, LDA gives a better result than eigenfaces, but it does not give the robust solution as their separable criterion is not relevant to classification precision [12]. Wavelet transform [5] is very good tool to analyze texture pattern in time and frequency domain, and these techniques work well for frontal faces only. But they are not robust against rotation variations because the whole-face-based process is highly sensitive toward translations and rotations. Another limitation of these approaches is that the size of the feature vector is too large to recognize. So these methods are computationally expensive.

Generally, a face image is represented by large number of features using many feature extraction methods. Some of the features have more discriminating power than the others. In other words, all features do not contribute equally to the face recognition process. It is not always true that the higher number of features lead to higher recognition rate. So optimum feature selection is a big issue in pattern recognition domain. Optimum feature selection has many advantages. It reduces the feature size and increases the recognition rate. Feature selection process first identifies the irrelevant features, discards them, and takes others which are treated as optimum features. Thus, feature selection is basically a search process. There are different search algorithms such as greedy [13], branch and bound [14], sequential search algorithms [15], mutual information [16], and tabu search [17] which have been used successfully. But these algorithms are quite computationally expensive. There are other kinds of population-based search algorithms in the literature, which is less expensive in times. Such algorithms are Genetic Algorithm (GA)-based method [18, 19, 20], and ant colony optimization (ACO)-based techniques have attracted a lot of attention [21] to the researchers.

In this paper, we propose an efficient scheme for scale and rotation invariant thermal face recognition using scale invariant feature transform (SIFT). The main contributions of this work are as follows:

- A complete scale and rotation invariant thermal face recognition system based on SIFT features are implemented.
- PSO-based feature selection algorithm is developed to search for the optimal features and to increase the recognition rate, as well.
- Evaluation of the proposed system using the UGC-JU thermal face database and comparing its performance with other FR techniques.

The outline of this paper is organized as follows: Sect. 2 describes the different steps of proposed approach including image preprocessing, features extraction, selection of features, and classification. The experiment and results are presented in Sect. 3. Finally, Sect. 4 concludes and remarks about some of the aspects analyzed in this paper.

2 Proposed Method

In this section, we have introduced a robust method for face recognition using thermal face images. The overall system includes image preprocessing, extraction of features, feature selection, and classification. All of these elements are explained in the following sections.

2.1 Preprocessing

Preprocessing is the first step of the proposed system. In this step, an intermediate image is produced from the 'just captured raw' image. A typical 24-bit color thermal face image is shown in Fig. 1. Here, each 24-bit color images have been converted into its corresponding grayscale images. Then, those converted grayscale images are again converted into binary images counterparts. After conversion of the binary image, it has been found that some white segments are present with a larger one in the binary image and this biggest part is the face area. So the largest part has been extracted from the binary image using connected component labeling algorithm [6], and other small components, which are other parts of the binary image, have been excluded. It has also been seen that some holes are created in the face area due to uneven distribution of thermal information which is nothing but temperature statistics. These temperature statistics has been excluded in the binarization process, which will make the face recognition process tougher.

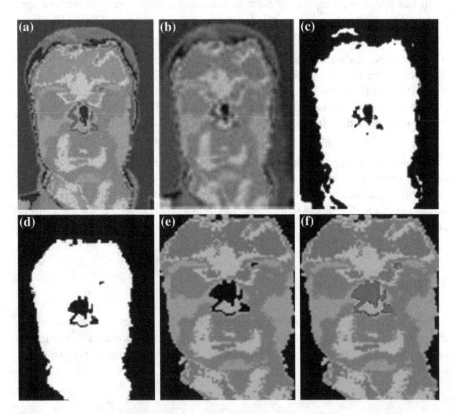

Fig. 1 The various outcomes of the preprocessing step. **a** A thermal face image. **b** Grayscale image. **c** Binary image. **d** Largest component as a face skin area. **e** Extracted face skin area in *gray* level. **f** Restored image

Principal component analysis (PCA) is reported to be robust for the problem of contaminated pixels [7]. But it cannot restore the lost information. So GPCA [8, 9] method is used to store the missing temperature statistic information. The outcomes of the preprocessing step are shown in Fig. 1.

2.2 Extraction of Features

The performance of any face recognition system highly depends on the selected features. So a good feature extraction algorithm is very necessary for face recognition which helps to identify probably the best discriminating power and which are less sensitive to variations in pose, scale, and illumination, and facial expressions etc. SIFT [10] is one such algorithm. SIFT is used for extraction of distinctive invariant features from the objects, which can be used to carry out the matching process. The features obtained by SIFT are invariant to object rotation and scale. It reduces the probability of reduced extraction due to occlusion and noise. First Gaussian function, $G(x, y, \sigma)$, is used to convolve with an input image $I(x, y)$ in order to get a scale-space image, $L(x, y, \sigma)$ image. The Gaussian function and scale-space image are found by the Eq. (1) and (2), respectively.

$$G(x, y, \sigma) = \frac{1}{2\pi\sigma^2} e^{\left(-\frac{x^2+y^2}{2\sigma^2}\right)} \tag{1}$$

$$L(x, y, \sigma) = G(x, y, \sigma) * I(x, y) \tag{2}$$

where '*' is the convolution operation in the x and y directions. So the initial image is incrementally convolved with Gaussians to produce images separated by a constant factor k in scale-space. Here, each scale-space is divided into 's' equal intervals, where 's' is a natural number and hence $k = 2^{1/s}$. Then, difference-of-Gaussian function convolved with the image, $D(x, y, \sigma)$, that can be computed from the difference of two nearby scales separated by a constant multiplicative factor k is calculated for the detection of efficient, stable landmarks. The difference-of-Gaussian function is shown in Eq. (3).

$$D(x, y, \sigma) = L(x, y, k\sigma) - L(x, y, \sigma) = (G(x, y, k\sigma) - G(x, y, \sigma)) * I(x, y) \tag{3}$$

Now, local maxima and local minima have been found from the image, $D(x, y, \sigma)$. First, a sample point from the current image has been chosen and compared to its eight neighbors in the current image and nine neighbors in the scale image above and below. It is chosen as a landmark only if it is larger (local maxima) than all of these neighbors or smaller (local minima) than all of them.

2.3 Feature Selection

After extraction of features, it has found that, within the extracted features, there are some features, which are irrelevant and noisy. These irrelevant and noisy features lead the misclassification rate. So the objective of feature selection step is to reduce the noisy data and exclude the irrelevant features as much as possible. In other word, find the optimal features from the original features including noisy and irrelevant features, which have higher discriminating power, to improve the recognition rate. Particle swarm optimization (PSO) is one such well-known tool to find the optimum characteristics with the help of local as well as global search in the feature search space in an iterative way. PSO proposed by Dr. Eberhart and Dr. Kennedy in 1995 [11]. In PSO, swarm consists of a group of random particles, which move around the solution space of the problem by updating through iterations for an optimum solution and go until convergence is achieved. A flowchart of the PSO-based system is given below:

Flowchart 1. PSO based feature selection

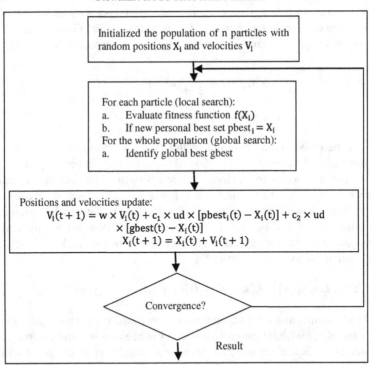

Initialized the population of n particles with random positions X_i and velocities V_i

For each particle (local search):
a.   Evaluate fitness function $f(X_i)$
b.   If new personal best set $pbest_i = X_i$
For the whole population (global search):
a.   Identify global best gbest

Positions and velocities update:
$$V_i(t+1) = w \times V_i(t) + c_1 \times ud \times [pbest_i(t) - X_i(t)] + c_2 \times ud \times [gbest(t) - X_i(t)]$$
$$X_i(t+1) = X_i(t) + V_i(t+1)$$

Convergence?

Result

In this work, 'n' number of random particles is chosen initially from the features space. Each particle having c parameters that are obtained, after feature extraction using SIFT operator, and their corresponding random velocities form a position matrix $X[n, c]$. Now, the threshold should be selected for the first round of selection of these random velocities and its corresponding positions by the following functions $V[i, j] = e(X[i, j])$ where $1 \leq i \leq n$ and $1 \leq j \leq c$ and it is assumed to be 0.5 for this work. The velocity of the ith particle is described by the $V_i = (v_{i1}, v_{i2}, \ldots, v_{ic})$, and its corresponding state is represented by $X_i = (x_{i1}, x_{i2}, \ldots, x_{ic})$. If the newly computed velocity is greater than the threshold value (0.5), then this velocity and its location is selected for the next iteration. It is expected that, after each iteration, the recognition rate of the face recognition system increases with the newly selected features from the features space. So the success rate is calculated by an objective function known as the fitness function in PSO. The minimum distance function [5] is used here as a fitness function for this work. Here, minimum distance classifier concentrates both local and as well as global information of the features obtained from SIFT operator. The fitness function is evaluated for each particle in the swarm and is compared to the fitness of the best previous (pbest) result for that particle and to the fitness of the best particle (gbest) among all particles in the swarm. After finding the two best values (pbest and gbest), the particles start updating their velocities and positions according to the Eqs. (4) and (5), respectively.

$$V_i(t + 1) = w \times V_i(t) + c_1 \times ud \times [\text{pbest}_i(t) - X_i(t)] + c_2 \times ud \times [\text{gbest}(t) - X_i(t)] \tag{4}$$

$$X_i(t + 1) = X_i(t) + V_i(t + 1) \tag{5}$$

where 'i' $= 1, \ldots, n$ and 'n' is the population size, 'ud' is another random number between 0 and 1, 'c_1' and 'c_2' are cognitive and social parameters, respectively, bounded between 0 and 1. In the velocity update equation, the + sign divides the whole equation into three components named as inertial component, a cognitive component, and social component, respectively. The inertia weight w is a factor used to control the balance of the search algorithm between exploration (=0.15) and exploitation (=1); the second element is the 'cognitive' section representing the local knowledge of the particle itself; the third component is the 'social' part, representing the cooperation among the particles. The iterative steps will go on until the process reaches the termination condition. It is experimentally found that thirty iterations are well enough to identify the optimum features from the features space, which leads the success rate to a great extent.

3 Experiment and Results

The performance of the proposed system is evaluated on newly created UGC-JU thermal face database [9]. UGC-JU thermal face database consists of 84 subjects each is having 39 different face images with Exp1 (happy), Exp2 (angry), Exp3 (sad), Exp4 (disgusted), Exp5 (neutral), Exp6 (fearful), and Exp7 (surprised) are considered. Different pose changes about x-axis, y-axis, and z-axis are also

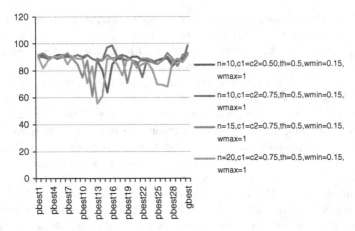

Fig. 2 The different values of 'pbest' and 'gbest' with different parameters of PSO for frontal faces

considered. Resolution of each image is 320 × 240, and the images are saved in JPEG format. First, all are processed and restored missing information using GPCA scale to 256 × 256 resolution for further processing. Then, SIFT is used to extract some points. After extraction of some points, total image is divided into several blocks of size 16 × 16 and find the number of points in each block and store them in a row vector. Hence, one row corresponds to one image and column corresponds to feature. So total number of features of a particular face or a row is (256 × 256)/ (16 × 16) = 256. But the 256 extracted features contain some noisy and irrelevant features. As a result, the performance of the system degrades. So selection of features which are relevant, free from noise, and redundancy is essential to improve the performance of the thermal face recognition system. PSO is used to find the optimal features. Ten optimum features are taken from 256 features of a face image using PSO so that they have enough discriminating power. Within this feature selection step, minimum distance classifier is used as a fitness function which identifies the class label of the testing images in successive iterations. Total four sets of experiments have been performed. First set of experiments has been performed on frontal thermal face images which includes various facial expressions. In this experiment, features coming from SIFT operator are considered and PSO is not used. In the second set of experiments, all thermal face images including pose changes about x-axis, y-axis, and z-axis, and occlusion with the frontal thermal face images are considered, and the features selection tool is not used. So, in the above two cases, the size of the features of each face image is 256. In the third set of experiment, SIFT and PSO are used for feature extraction and feature selection, respectively, on the frontal thermal face images only. Different parameters for PSO like number of population 'n,' c_1, c_2 are varied, and the obtained results are shown in Fig. 2. The x-axis of this figure is represented by 'pbest' in each iteration, 'gbest' and the y-axis demonstrated the percentage of recognition rate in each iteration. Finally, SIFT and PSO are used on all thermal face images. In all these cases, total

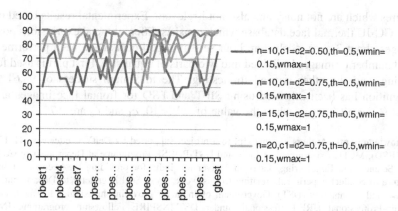

Fig. 3 The different values of 'pbest' and 'gbest' with different parameters of PSO for all faces

Table 1 The recognition rate in percentage

Method	Frontal thermal face images	All thermal face images
SIFT only	94.44	86.84
SIFT + PSO	98.61 (gbest)	90.28 (gbest)
Culter [2]	72.22	70.83

images are divided into two sets. First set is used for training, and other is used for testing purpose. Thus, from the Figs. 2 and 3, it is clear that the 'gbest' means global best recognition is 98.61 and 90.28 % when the value of 'n' is 10, c_1 and c_2 are 0.75 for frontal faces and all face images, respectively.

3.1 Comparison with Other Methods

The obtained recognition rate by the present method has been compared with one pioneer work on recognition of infrared face images by Culter [2]. The work has been implemented on UGC-JU thermal face images. The obtained results are shown in Table 1. All the results support the conclusion that the face recognition performances by the present method give better results in frontal images as well as all the images.

4 Conclusions

This paper presents a scale, translation, and rotation invariant thermal face recognition system using SIFT for face recognition systems. In this system, a PSO-based feature selection algorithm is efficiently utilized to search the optimum features which are not noisy and also not irrelevant. Experimental results based on the UGC-JU thermal face database which consists different facial expressions pose changes about 3 axes, occlusion. Four sets of experiments have been performed. Total number of images is divided into two parts of equal size. First part is used for training, and the other is used for testing. The maximum success of 98.61 % recognition has been achieved using SIFT and PSO for frontal face images and 90.28 % for all images, when the value of 'n' is 10, c_1 and c_2 are 0.75.

Acknowledgments Authors are thankful to a project supported by DeitY (Letter No.: 12(12)/2012-ESD), MCIT, Government of India and DST-PURSE Programme at Department of Computer Science and Engineering, Jadavpur University, India for providing the necessary infrastructure to conduct experiments relating to this work. Ayan Seal is grateful to Department of Science and Technology (DST), Government of India for providing him Junior Research Fellowship-Professional (JRF-Professional) under DST-INSPIRE Fellowship programme [No: IF110591]. Ayan Seal is also thankful to Universidad Politecnica de Madrid, Spain, for providing him scholarship under Erasmus Mundus Action 2 India4EU II.

References

1. Kong, S.G., Heo, J., Abidi, B.R., Paik, J., Abidi, M.A.: Recent advances in visual and infrared face recognition—a review. Comput. Vis. Image Underst. **97**, 103–135 (2005)
2. Cutler, R.: Face recognition using infrared images and eigenfaces. University of Maryland, Technology Report, College Park (1996)
3. Turk, M., Pentland, A.: Eigenfaces for recognition. J. Cogn. Neurosci. **3**(1), 71–86 (1991)
4. Fisher, R.A.: The statistical utilization of multiple measurements. Ann Eugenics **8**, 376–386 (1938)
5. Bhattacharjee, D., Seal, A., Ganguly, S., Nasipuri, M., Basu, D.K.: A comparative study of human thermal face recognition based on haar wavelet transform and local binary pattern. Comput. Intell. Neurosci. doi:10.1155/2012/2610892012
6. Morse, B.S.: Lecture 2: Image Processing Review, Neighbors, Connected Components, and Distance (1998–2004)
7. Leonardis, A., Bischof, H.: Robust recognition using eigenimages. Comput. Vis. Image Underst. **78**(1), 99–118 (2000)
8. Everson, R., Sirovich, L.: Karhunen: Loeve procedure for gappy data. J. Opt. Soc. Am. A **12**(8), 1657–1664 (1995)
9. Seal, A., Bhattacharjee, D., Nasipuri, M., Basu, D.K.: UGC-JU face database and its benchmarking using linear regression classifier. Multimedia Tools Appl. (2013). doi:10.1007/s11042-013-1754-8
10. Lowe, D.: Distinctive image features from scale-invariant keypoints. Int. J. Comput. Vis. **60**, 91–110 (2004)
11. Eberhart, R.C., Kennedy, J.: A new optimizer using particle swarm theory. In: Proceedings of the 6th International Symposium on Micromachine and Human Science, Nagoya, Japan, pp. 39–43 (1995)
12. Ramadan, R.M., Abdel-Kader, R.F.: Face recognition using particle swarm optimization-based selected features. Int. J. Signal Process. Image Process. Pattern Recogn. **2**(2), 51–65 (2009)
13. Kokiopoulou, E., Frossard, P.: Classification-specific feature sampling for face recognition. In: Proceedings of the IEEE 8th Workshop on Multimedia Signal Processing, pp. 20–23 (2006)

14. Narendra, P.M., Fukunage, K.: A branch and bound algorithm for feature subset selection. IEEE Trans. Comput. **6**(9), 917–922 (1977)
15. Pudil, P., Novovicova, J., Kittler, J.: Floating search methods in feature selection. Pattern Recogn. Lett. **15**, 1119–1125 (1994)
16. Roberto, B.: Using mutual information for selecting features in supervised neural net learning. IEEE Trans. Neural Networks **5**(4), 537–550 (1994)
17. Zhang, H., Sun, G.: Feature selection using Tabu search method. Pattern Recogn. Lett. **35**, 701–711 (2002)
18. Kim, D.-S., Jeon, I.-J., Lee, S.-Y., Rhee, P.-K., Chung, D.-J.: Embedded face recognition based on fast Genetic Algorithm for intelligent digital photography. IEEE Trans. Consum. Electron. **52**(3), 726–734 (2006)
19. Raymer, M.L., Punch, W.F., Goodman, E.D., Kuhn, L.A., Jain, A.K.: Dimensionality reduction using Genetic Algorithms. IEEE Trans. Evol. Comput. **4**(2), 164–171 (2000)
20. Fan, X., Verma, B.: Face recognition: a new feature selection and classification technique. In: Proceedings of the 7th Asia-Pacific Conference on Complex Systems, (2004)
21. Kanan, H.R., Faez, K., Hosseinzadeh, M.: Face recognition system using ant colony optimization-based selected features. In: Proceedings of the IEEE Symposium Computational Intelligence in Security and Defense Applications (CISDA 2007), pp. 57–62 (2007)
22. Seal, A., Bhattacharjee, D., Nasipuri, M., Basu, D.K.: UGC-JU Face Database and its Benchmarking using Linear Regression Classifier, Multimedia Tools and Applications, Springer, US, 10.1007/s11042-013-1754-8, (2013)

Retinal Feature Extraction with the Influence of Its Diseases on the Results

Anna Bartocha, Emil Saeed, Piotr Wachulec and Khalid Saeed

Abstract A new algorithm on retinal characteristics extraction is introduced in this paper. The important thing about its novelty is that it considers and treats eyes with anomalies. The background of both medical and computer science matters is given. The algorithm aims at giving a clear evidence whether the retina should be considered as a biometric feature in human recognition for people with sick eyes. The structure of the worked out algorithm is illustrated in detail. Examples of how the minutiae are extracted from the processed retina are presented. The algorithm details together with its different stages, and their computer implementation will be given in the extended version of the work.

Keywords Biometrics · Human identification · Identification of retina with anomalies · Retinal diseases

A. Bartocha (✉) · P. Wachulec · K. Saeed
Faculty of Physics and Applied Computer Science, AGH University of Science and Technology, Krakow, Poland
e-mail: bartochaanna@gmail.com

P. Wachulec
e-mail: pioxan@gmail.com

K. Saeed
e-mail: saeed@agh.edu.pl

E. Saeed
Faculty of Medicine, Department of Ophthalmology, Medical University of Bialystok, Bialystok, Poland
e-mail: emilsaeed1986@gmail.com

K. Saeed
Faculty of Computer Science, Bialystok University of Technology, Bialystok, Poland

© Springer India 2015

R. Chaki et al. (eds.), *Applied Computation and Security Systems*, Advances in Intelligent Systems and Computing 304, DOI 10.1007/978-81-322-1985-9_3

1 Introduction

The retina is a highly complex structure that consists of many layers. Optic disc nerve is a location on the retina where the optic nerve exits the eye. The optic disc is also the entry point for the major blood vessels that supply the retina.

The reason why we use retinal blood vessels to recognize the person is that human retina is unique [1]. Even identical twins do not share a similar pattern of the blood vessels network in the retina.

However, there still are some disorders in these vessels caused usually by serious diseases, which may prevent the pattern identification. Here are some examples of such cases.

1. Central retinal artery occlusion, where a part of the artery may be unseen.

Figure 1 shows this concept.
Central retinal artery occlusion is a disease where blood flow through retinal artery is blocked. The patient complains of sudden painless loss of vision. Cholesterol and calcific emboli may result in permanent obstruction. Sometimes the artery can recapitalize with time.

2. Central retinal vein occlusion

Central retinal vein occlusion may result from abnormality of blood itself, an inflammation or an increased ocular pressure. The patient complains of sudden partial or complete loss of vision. We can observe a haemorrhage and swelling and tortuosity of the veins. It can recapitalize with time. There are some risk factors for the development of this disease, for example hypertension, obesity, diabetes or glaucoma.

3. Retinal detachment, which may hide the retinal vessels (Fig. 2).

There are many potential causes or risk factors: head damages, diabetic retinopathy, diabetic retinopathy, myopia (short-sighted) inflammations and tumours. Patient notices the progressive development of a field defect, described as a 'shadow' or 'curtain'. It can be treated surgically.

2 Proposed Methodology

This section presents the proposed methodology for retinal image processing and recognition. Example of retinal image of healthy person is shown in Fig. 3.

The main idea of this algorithm is to create a vector of characteristic points at retinal image. We can compare these points to decide whether two pictures present the same person or not. The characteristics can be minutiae of retinal vessels and its location. The landmarks of retinal vessels are bifurcations, crossings and end points. General parts of retinal image processing at proposed approach are shown in Fig. 4.

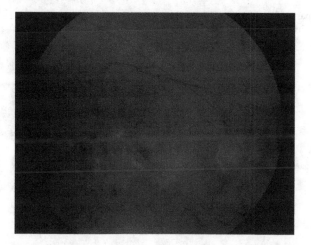

Fig. 1 Retinal occlusion

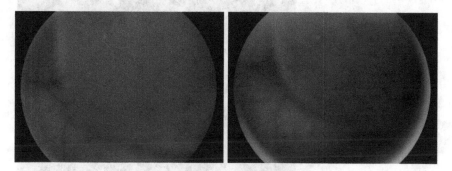

Fig. 2 Retinal detachment—retina peels away from the underlying layer

Fig. 3 Example of retinal image

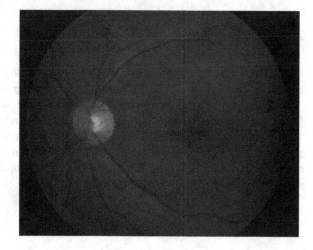

Fig. 4 Basic parts of the suggested algorithm

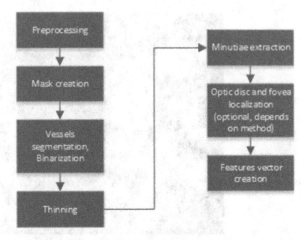

Fig. 5 Image after preprocessing

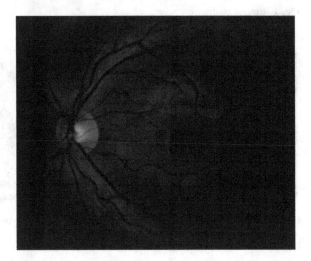

2.1 Preprocessing

First, image should be preprocessed. If it is necessary, we can do denoising, contrast enhancement, convert to greyscale and histogram normalization.

It is noticed that retinal vessel is the best visible on green colour channel. Because of this fact, we can use only that layer for searching of minutiae (Fig. 5).

We applied two methods to feature extraction and compared results. First method is based on greyscale calculated as average of red, green and blue channel of image. Second method use only green channel. Comparison and results of both methods are presented in results section.

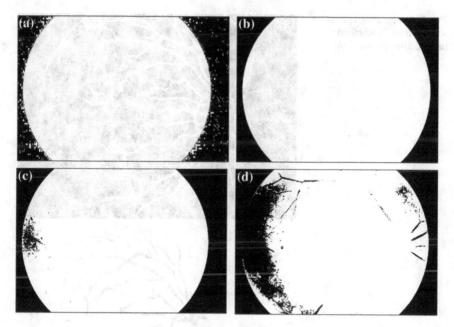

Fig. 6 Masks with threshold value 1 (**a**), 20 (**b**), 40 (**c**) and 50 (**d**)

2.2 Mask Creation

We have to create a mask, because we do not want to take into consideration black frame of the image. The best way to do this is by using thresholding with very low parameter. Additionally, we can remove small artefacts, such as some noise in the middle of image.

Results with different values of threshold are presented below. If threshold is too high, there is unnecessary noise inside the mask. But if this value is too low, there are white spots in mask (Fig. 6).

2.3 Vessels Segmentation

The next step is vessel enhancement and segmentation. The simplest solution could be proper binarization, but there are no good results then. One of the approach is using Gaussian matched filter [2] and then binarization (for example local entropy thresholding [3, 4]). Blood vessels usually have poor local contrast, and edge detection algorithms results are not sufficient. In Gaussian matched filter method, we receive greyscale image approximated by Gaussian-shaped curve. The aim is to detect piecewise linear segments of blood vessels so we create 12 different mask filters and search for vessels in every 15°. Figure 7 illustrates this step.

Fig. 7 Image after vessels segmentation: uncleaned (**a**) and after cleaning (**b**)

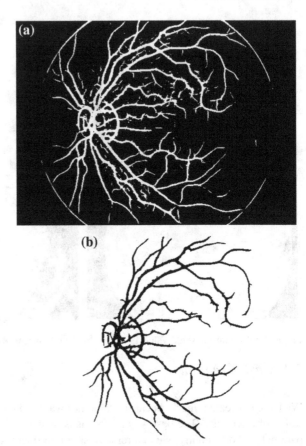

2.4 Thinning

When we have black and white image of vessels structure, we can do thinning. This operation allows us to get backbone of image. It can be used, for example mask thinning and algorithms such as KMM [5] or K3M [6] (Fig. 8).

2.5 Minutiae Extraction

Next step is finding bifurcations, crossings and end points of vessels. It can be done with the aid of masks. The minutiae and the required masks are shown in Fig. 9.

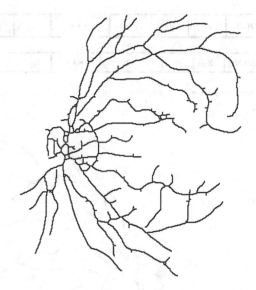

Fig. 8 Image after K3M thinning

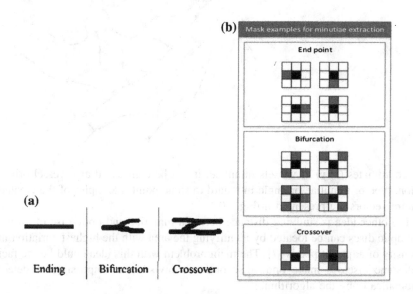

Fig. 9 Types of minutiae (**a**) and their mask examples (**b**)

2.6 Feature Vector Creation

At the end, the vector of characteristics should be created. We have to decide which data will describe the image of Fig. 11. One of the approaches is that there

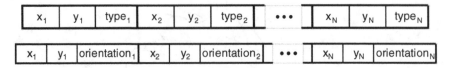

| x_1 | y_1 | $type_1$ | x_2 | y_2 | $type_2$ | $\bullet\bullet\bullet$ | x_N | y_N | $type_N$ |

| x_1 | y_1 | $orientation_1$ | x_2 | y_2 | $orientation_2$ | $\bullet\bullet\bullet$ | x_N | y_N | $orientation_N$ |

Fig. 10 Examples of feature vectors

Fig. 11 Retinal image after finding minutiae

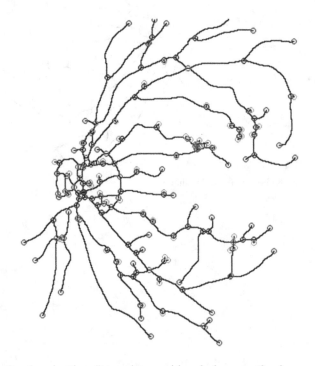

are coordinates (x, y) of vessels minutiae. It can be considered also vessel orientation, type of minutiae or angle in regard to some point. Examples of the created feature vectors are presented in Fig. 10.

The other idea is studying distance between minutiae and optic disc of retina. The optic discs can be located by identifying the area with the highest variation in intensity of adjacent pixels [7]. The main problem with this idea could be the fact that some diseases make optic disc or fovea invisible or impossible to detect automatically by the algorithm.

This feature vector will be used later in classification process.

Next step of biometrics system will be verification. There will be some kind of database with retinal feature vectors. We acquire retinal picture of person who will be investigated. Then, we calculate distance between feature vectors from database and this one from person being tested. We search for the shortest distance and define probability if he is the person he really is.

3 Experimental Results

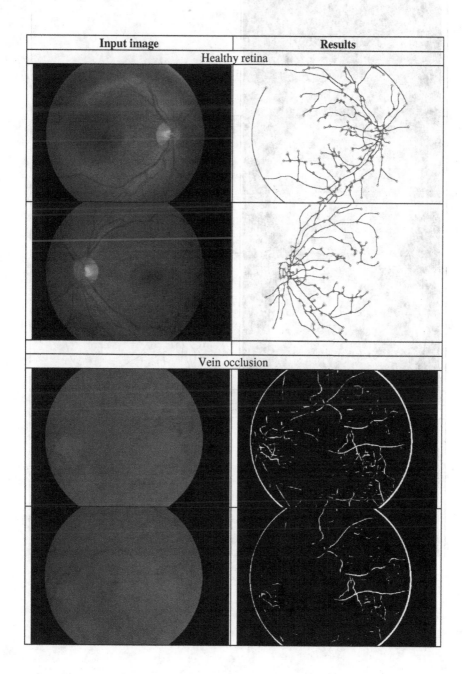

Input image	Results
Healthy retina	
Vein occlusion	

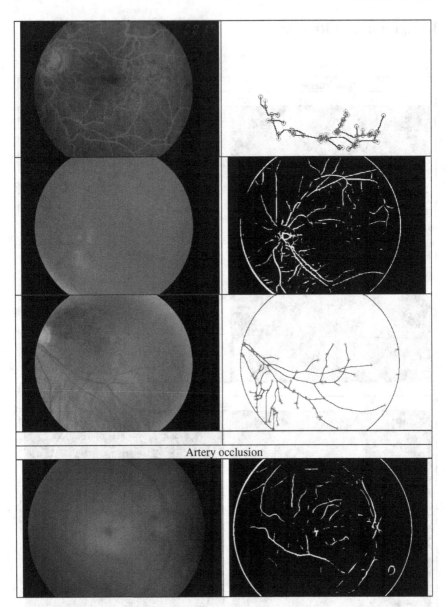

Artery occlusion

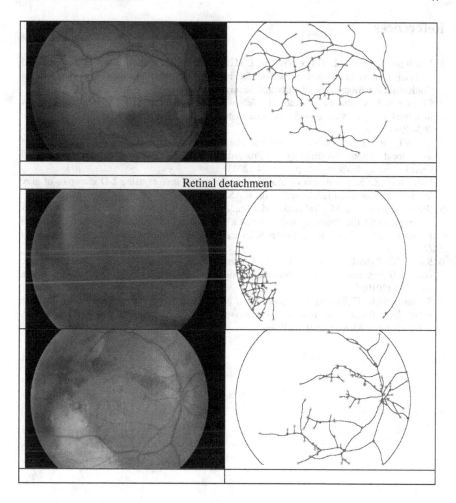

Retinal detachment

4 Result Analysis and Conclusions

As we can see, there is difference in points (quantitative and locally) in case of healthy and diseased eye. Vein occlusion and artery occlusion often result in dashed veins after usage of Gaussian matched filter. Current version of the algorithm cannot recover from that state, so large part of information is lost by cleaning step which proceeds thinning.

It is expected that this loss will impact on effectiveness of algorithm. This requires further research. A research team is working on more cases and classification of results in order for people verification.

Acknowledgments The research was partially supported by Grant No. WFiIS 11.11.220.01/ saeed, AGH University of Science and Technology in Krakow.

References

1. Villalobos-Castaldi, M., Felipe-Riverón, E.M.,: Fast automatic retinal vessel segmentation and vascular landmarks extraction method for biometric applications. In: IEEE International Conference on Biometrics, Identity and Security (2009)
2. Chaudhuri, S., Chatterjee, S., Katz, N., Nelson, M., Goldbaum, M.: Detection of blood vessels in retinal images using two-dimensional matched filters. IEEE Trans. Med. Imaging 8(3), 263–269 (1989)
3. Chanwimaluang, T., Fan, G.: An efficient blood vessel detection algorithm for retinal images using local entropy thresholding. In: Proceedings of the 2003 International Symposium on Circuits and Systems, vol. 5, pp. 21–24 (2003)
4. HongQing, Z.: Segmentation of blood vessels in retinal images using 2-D entropies of gray level-gradient co-occurrence matrix. In: ICASSP (2004)
5. Saeed, K., Rybnik, M., Tabedzki, M.: Implementation and advanced results on the non-interrupted skeletonization algorithm. In: Skarbek, W. (ed.) Computer Analysis of Images and Patterns, Lecture Notes in Computer Science, vol. 2124, Springer, Heidelberg, pp. 601–609 (2001)
6. Saeed, K., Rybnik, M., Tabedzki, M., Adamski, M.: K3M: a universal algorithm for image skeletonization and review of thinning techniques. Int. J. Appl. Math. Comput. Sci. 20(2), 317–335 (2010)
7. Sinthanayothin, C., Boyce, J.F., Cook, H.L., Williamson, T.H.: Automated localisation of the optic disc, fovea, and retinal blood vessels from digital colour fundus images. Br. J. Ophthalmol. 83(8), 902–910 (1999)

Iris Feature Extraction with the Influence of Its Diseases on the Results

Piotr Wachulec, Emil Saeed, Anna Bartocha and Khalid Saeed

Abstract An algorithm for human iris recognition is presented in this paper. The essential idea of the work is to show the results of the authors' investigation in treating sick eyes with iris anomalies. This is actually specific as the iris code may (or may not) be affected in such situations. The paper introduces the basic steps in iris image and its main characteristics and features extraction leaving the detailed description of the algorithm and its results to the extended version of the paper. The answer to the main problem of the investigation is supposed to be given or at least discussed to know the relation between eye iris of sick and healthy eyes.

Keywords Biometrics · Human identification · Iris recognition · Identification of iris with anomalies · Iris diseases

P. Wachulec (✉) · A. Bartocha · K. Saeed
Faculty of Physics and Applied Computer Science, AGH University of Science and Technology, Krakow, Poland
e-mail: pioxan@gmail.com

A. Bartocha
e-mail: bartochaanna@gmail.com

K. Saeed
e-mail: saeed@agh.edu.pl

E. Saeed
Department of Ophthalmology, Faculty of Medicine, Medical University of Bialystok, Bialystok, Poland
e-mail: emilsaeed1986@gmail.com

K. Saeed
Faculty of Computer Science, Bialystok University of Technology, Bialystok, Poland

© Springer India 2015
R. Chaki et al. (eds.), *Applied Computation and Security Systems*, Advances in Intelligent Systems and Computing 304, DOI 10.1007/978-81-322-1985-9_4

1 Introduction

The iris is the colored structure of the eye that divides the space between the cornea and the lens into anterior and posterior chambers. It is built from two pigmented layers—stroma and the pigmented cells (for the anatomy details and pictures see [1], Chap. 3).

The complicated structure of the iris and the fact that there are no two identical irises (even the left and right eyes of the same person) shows why we use it for personal identification.

However, there are some anomalies or diseases that may change or hide the structure of the iris, and hence, they may affect the iris recognition. Although many these anomalies are known to us, there still are some odd cases that need special attention and deep study. The iris images shown in Figs. 1 and 2 are examples.

The question that will arise is whether the image after processing would keep the characteristics of the iris or not.

There are many articles describing iris recognition problem. The most popular are articles written by Daugman [2–5]. He proposed to represent iris as a mathematical function. He used integro-differential operator, neural network, 2D Gabor transform, and Hamming distance. Another example is work by Wildes [6] who used Hough transform and Laplacian of Gaussian filters. Then, Boles and Boashash [7] with wavelet transform zero crossing and dissimilarity function, but it was not perform for non-ideal iris images. Next, Poursaberi and Araabi [8] worked out method with morphological operators and thresholds, Wiener 2D filter, and Daubechies 2 wavelet.

2 Proposed Methodology

This section presents the proposed methodology for iris image processing and recognition. Figure 3 shows the block diagram of the suggested algorithm in its essential stages.

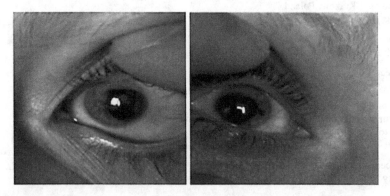

Fig. 1 A congenital syndrome called Axenfeld-Rieger syndrome, with the absence in some parts of the iris

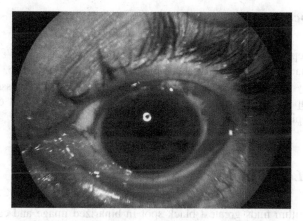

Fig. 2 Hyphema—blood in anterior chamber and laser iridotomy, which is a small hole as a treatment in glaucoma

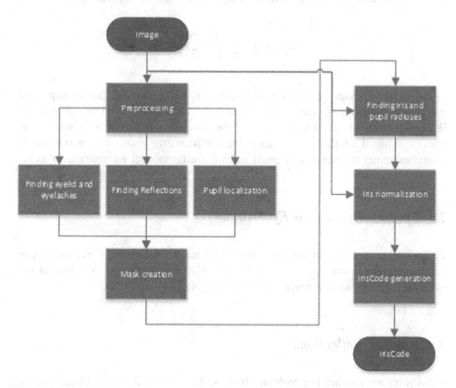

Fig. 3 General algorithm flow

2.1 Preprocessing

In the first step, we need to prepare images for further processing. We would like to achieve that by removing artifacts from camera. In the next step, we convert image to gray scale. Furthermore, we can smooth image with Gaussian filter to offset the impact of eyelashes. Also we need to prepare binarized copy of the image for pupil localization.

2.2 Pupil Localization

Current algorithm finds greatest black spot in binarized image and calculates its dimensions and position. With this data, it is possible to find pupil center as the average of the coordinates. Unfortunately, this method has a lot of disadvantages, so we would like to develop method using the differential-integral operator:

$$\max_{r,x_0,y_0} \left| G_\sigma(r) \cdot \frac{\partial}{\partial r} \oint_{r,x_0,y_0} \frac{I(x,y)}{2\pi r} ds \right|$$

I(x, y) is the image in spatial coordinates, r is the radius, (x_0, y_0) are pupil and iris center coordinates, and $G\sigma(r)$ is Gaussian smoothing function of scale σ. Integro-differential operator searches got the maximum of partial derivative of the contour integral of the image along the circular arc. Using that maximum it estimates center coordinates of pupil and iris furthermore it estimates their radii.

2.3 Finding Eyelid and Eyelashes

To find the parts of the image containing the iris data, we should find positions of the eyelids, which can cover part of the iris. Also we should find positions of the eyelashes which can add irrelevant noise (Fig. 4).

2.4 Finding Reflections

Reflections are another big problem, because they can remove significant part of iris data. Currently, algorithm does not have any method to deal with them; however, we would like to create method which would help us determine their positions and shape to create mask for next step. Figure 5 shows the possible reflection existence in the eye image.

(a)   **(b)**

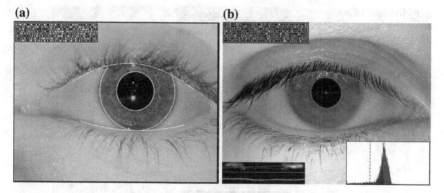

Fig. 4 Detected iris and eyelids [9] (**a**) and the eyelash detection [10] (**b**)

Fig. 5 Example of possible reflections and other artifacts [11]

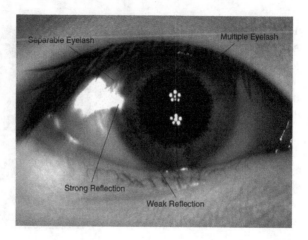

2.5 Mask Creation

After localization of eyelids, eyelashes, and reflections, it will be possible to create mask that will allow us to remove all unwanted artifacts from the image for further processing.

2.6 Finding Iris and Pupil Radius

In current version of the algorithm, iris and pupil are considered as circles with common center; however, real iris images do not have to be so ideal. Solution to this problem is development of active contour for both pupil and iris (Fig. 6).

Fig. 6 Example of active
contour [10]

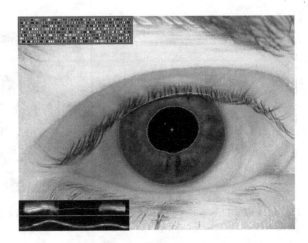

Fig. 7 Axenfeld-Rieger
syndrome

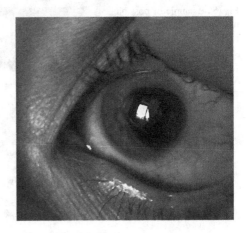

Only by using that method it would be possible to acquire normalized iris images of people with diseases such as Axenfeld-Rieger syndrome (Fig. 7).

2.7 Iris Normalization

The whole iris is normalized to rectangular area (Fig. 8). With simple circles, it is just simple mapping of polar coordinates. With active contour, the whole mapping will be much more complicated.

Fig. 8 Normalized iris image

Fig. 9 Demodulation
example [9]

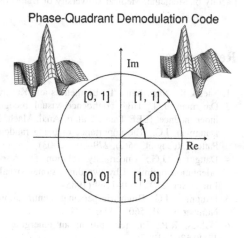

2.8 Features Vector Extraction

Features vector will be generated in form of IrisCode [9], as a result of writing real
and imaginary part of integral:

$$h_{\{Re, Im\}} = sgn_{\{Re, Im\}} \int_{\rho} \int_{\phi} I(\rho, \phi) e^{-i\omega(\theta_0 - \phi)} \cdot e^{-\frac{(r_0 - \rho)^2}{\alpha^2}} \cdot e^{-\frac{(\theta_0 - \phi)^2}{\beta^2}} \cdot \rho d\rho d\phi$$

We will use only phase bits acquired from the integral according to Fig. 9.
It will allow us to create 256-byte codes, which we can easily compare.

2.9 Feature Vectors Comparison

To compare 2048-bit codes, we need to calculate their Hamming distance using
masks which describes positions of artifacts in sample.

$$HD = \frac{\|(codeA \otimes codeB) \cap maskA \cap maskB\|}{\|maskA \cap maskB\|}$$

⊗ Boolean XOR operator
∩ Boolean AND operator

Acknowledgment The research was partially supported by grant no. WFiIS 11.11.220.01, AGH University of Science and Technology in Krakow and also by Department of Ophthalmology, Faculty of Medicine, Medical University of Bialystok.

References

1. Saeed, K., Nagashima, T.: Biometrics and Kansei Engineering. Springer, New York (2012)
2. Daugman, J.G.: High confidence visual recognition of persons by a test of statistical independence. IEEE Trans. Pattern Anal. Mach. Intell. **15**(11), 1148–1161 (1993)
3. Daugman, J.G.: The importance of being random: statistical principles of iris recognition. Pattern Recognit. **36**(2), 279–291 (2003)
4. Daugman, J.G.: Uncertainty relation for resolution in space, spatial frequency, and orientation optimized by two-dimensional visual cortical filters. J. Opt. Soc. Am. A Opt. Image Sci. **2**(7), 1160–1169 (1985)
5. Daugman, J.G.: Biometric personal identification system based on iris analysis. US Patent Number 5, 291, 560, 1 Mar 1994
6. Wildes, R.P.: Iris recognition: an emerging biometric technology. Proc. IEEE **85**(9), 1348–1363 (1997)
7. Boles, W.W., Boashash, B.: A human identification technique using images of the iris and wavelet transform. IEEE Trans. Signal Process. **46**(4), 1185–1188 (1998)
8. Poursaberi, A., Araabi, B.N.: Iris recognition for partially occluded images: methodology and sensitivity analysis. EURASIP J. Adv. Signal Process. **2007**(1), 20 (2007)
9. Daugman, J.: How iris recognition works. IEEE Trans. Circ. Syst. Video Technol. **14**, 1 (2004)
10. Daugman, J.: New methods in iris recognition. IEEE Trans. Syst., Man, Cybern.—Part B Cybern. **37**(5), 10 (2007)
11. Kong, W.K., Zhang, D.: Accurate Iris Segmentation Based on Novel Reflection and Eyelash Detection Model. In: Proceedings of 2001 International Symposium on Intelligent Multimedia, Video and Speech Processing (2001)

Part II
Imaging and Healthcare Applications

An Automatic Non-invasive System for Diagnosis of Tuberculosis

Pramit Ghosh, Debotosh Bhattacharjee and Mita Nasipuri

Abstract Tuberculosis (TB) is one of the major infectious diseases in the third world countries especially African countries. TB is a curable disease if proper treatment starts on time. Sometimes, it becomes difficult to arrange sufficient infrastructure for diagnosis of the disease in the remote areas of third world countries. Telemedicine partially solves the lack of physician and technicians in the pathological laboratories. The objective of this work is to design a complete setup that can be deployed at the remote areas of the country as a supporting aid of a telemedicine system for the diagnosis of TB. This system performs analysis of images of sputum and cough of patients. Microscopic image of TB bacteria which are stained with pink color is the main criterion for detection of TB disease. This system is still in development stage. A sufficient amount of data are needed for validation of the system.

Keywords Clustering · HSI color format · H-bridge · Sputum microscopy · Tuberculosis · Wiener filtering

P. Ghosh (✉)
Department of Computer Science and Engineering, RCC Institute of Information
Technology, Kolkata 700015, India
e-mail: pramitghosh2002@gmail.com

D. Bhattacharjee · M. Nasipuri
Department of Computer Science and Engineering, Jadavpur University,
Kolkata 700032, India
e-mail: debotoshb@hotmail.com

M. Nasipuri
e-mail: mitanasipuri@gmail.com

© Springer India 2015
R. Chaki et al. (eds.), *Applied Computation and Security Systems*, Advances in Intelligent
Systems and Computing 304, DOI 10.1007/978-81-322-1985-9_5

1 Introduction

Tuberculosis (TB) is an infectious disease that is caused by a bacterium called Mycobacterium TB [1]. TB primarily affects the lungs, but it can also affect organs in the central nervous system, lymphatic system, and circulatory system among others. TB is spread from person to person through the air. When people with lung TB cough, sneeze, or spit, they propel the TB germs into the air. A person needs to inhale only a few of these germs to become infected. About one-third of the world's population has latent TB, which means people have been infected by TB bacteria, but they are not yet sick with disease and cannot transmit the disease. People infected with TB bacteria have a lifetime risk of falling ill due to TB with a certainty of 10 %. However, persons with compromised immune systems, such as people living with HIV, malnutrition, or diabetes, or people who use tobacco, have a much higher risk of falling ill. When a person develops active TB disease, the symptoms like cough, fever, night sweats, weight loss, etc., may be mild for many months. This can lead to delay in seeking care, and results in transmission of the bacteria to others. People sick with TB can infect up to 10–15 other people through close contact over the course of a year. Without proper treatment, up to two-third of people, who are ill with TB, will die [2].

According to World Health Organization report of 2013 [3], an estimated 8.6 million people developed TB and 1.3 million died from the disease (including 320,000 deaths among HIV-positive people). The number of TB deaths is unacceptably large given that most are preventable. This report reveals that over 95 % of TB deaths occur in low- and middle-income countries, and it is among the top three causes of death for women aged 15–44. Furthermore, in 2012, an estimated 530,000 children became ill with TB and 74,000 HIV-negative children died of TB.

Sputum microscopy [4, 5] is one of the widely used noninvasive techniques to confirm a diagnosis of TB, but this technique is time-consuming because every sample slide is checked manually using microscope. Figure 1 shows the microscopic image of TB bacteria which are stained with pink color. To detect the disease, other methods like tuberculin skin test [6, 7], sputum culture [8], blood test, etc., require a long time, sometimes a week.

It will be very helpful for doctors if an automatic system compatible with telemedicine system [9] can be deployed at the remote places of the country to identify and classify the stain of TB from the cough samples. Moreover, it would be advantageous if only school education is sufficient to operate such diagnosis system in a remote place.

The objective of this work is to design a low-cost device which is capable of detecting and classifying the stains of TB from the microscopic images of cough samples to speed up the diagnosis process.

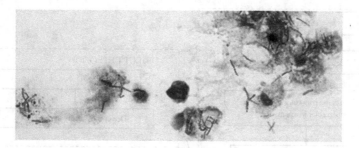

Fig. 1 Magnified image of the sample, with the presence of Mycobacterium tuberculosis

2 The System

The system is explained with the help of a block diagram, shown in Fig. 2, and all the steps of the system are described next.

2.1 Image Acquisition

Image acquisition is the first step of the system. Microscopic images of the blood samples of the slides are acquired with a CCD [10] camera which is mounted upon the eyepiece of a compound microscope. A fully automatic slide movement system is introduced to control the slide movement under the objective lens for different site images of the same sample slide. Two stepper motors are used for the sample slide movement in the X and Y directions shown in Fig. 3. Figure 4 shows the driving circuit of the stepper motor through parallel printer port (DB25) where L293D (H-bridge) IC is used for current amplification to control the stepper and apply voltage to each of the coils in a specific sequence. The sequence would go like Table 1.

2.2 Image Enhancement

The images obtained from the CCD camera are not up to the mark. Wiener filtering [11] is an effective linear image restoration approach, and it is used here. The task is to find the estimate of the 'best' image $\hat{f}$ and is done usually in frequency domain:

$$\hat{F}(u,v) = \frac{|H(u,v)|^2}{H(u,v)|H(u,v)|^2 + |N(u,v)|^2/|F(u,v)|^2} G(u,v) \qquad (1)$$

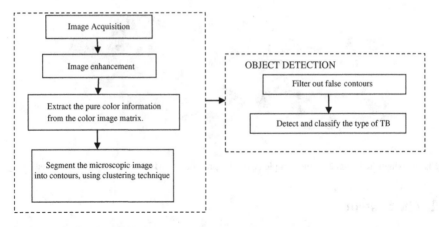

Fig. 2 Block diagram of the system

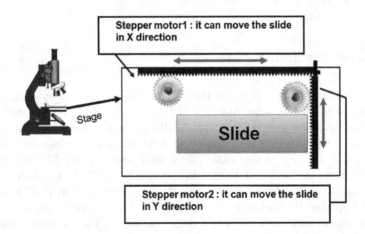

Fig. 3 Slide control using a stepper motor

where $H(u, v)$ is the degradation function, $G(u, v)$ is the Fourier transform of the degraded image, $N(u, v)2$ is the power spectrum of noise, and $F(u, v)$ 2 is the power spectrum of the undegraded image.

The Wiener filter performs deconvolution in the sense of minimizing a least squares error:

$$e^2 = E\left\{\left(f - \hat{f}\right)^2\right\} \tag{2}$$

Where E is the mean value and f is the undegraded image which is usually not known. In the case of omitting the noise, the Wiener filter changes to a simple

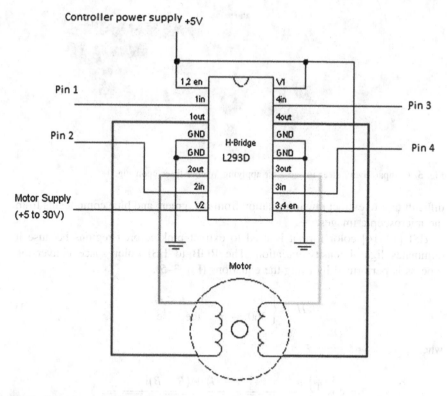

Fig. 4 Stepper motor driving circuit

Table 1 The pulse sequences to drive stepper motor

Step	Wire 1	Wire 2	Wire 3	Wire 4
1	High	Low	High	Low
2	Low	High	High	Low
3	Low	High	Low	High
4	High	Low	Low	High

inverse filter. The Wiener filter is applied separately on Red, Green, and Blue components of the color images obtained from the CCD camera. Figure 5 shows the image after applying Wiener filter.

2.3 Information Extraction

A huge number of pink threadlike elements are present in Figs. 1 and 5; actually, they are Mycobacterium TB bacteria. Due to variation in staining technique, the stained color is not always the same; it varies from reddish to pink. So, it is a very

After filtering

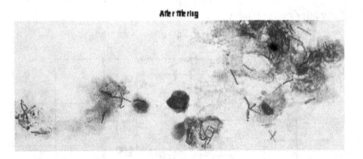

Fig. 5 Comparatively clear image after applying Wiener filter upon Fig. 1

difficult task to extract raw information from red, green, and blue components from the microscopic images.

HSI [12–14] color format is used to extract pink-colored regions because it eliminates light intensity variation. The RGB to HSI color space conversion process is performed by using the equations (Eq. 3–5).

$$H = \begin{cases} \theta & \text{if } B \leq G \\ 360° - \theta & \text{otherwise} \end{cases} \tag{3}$$

where

$$\theta = \cos^{-1}\left\{ \frac{\frac{1}{2}[(R - G) + (R - B)]}{\sqrt{[(R - G)(R - G) + (R - B)(G - B)]}} \right\}$$

$$S = 1 - \frac{3}{(R + G + B)}[\min(R, G, B)] \tag{4}$$

$$I = \frac{1}{3}(R + G + B) \tag{5}$$

where H stands for Hue, i.e., pure color, S for saturation, i.e., the degree by which the pure color is diluted using white light, and I for intensity, i.e., gray level.

2.4 Segmentation

By analyzing Hue components obtained from Eq. 3, it is inferred that the area of the image occupied by the TB bacteria has a HUE value very close to the HUE value of pink and high intensity variation with respect to other area of the image. So, partial supervised clustering techniques [15, 16] are applied, where initial cluster centers are the Hue value of pink, blue, white, and violet. Algorithm 1 is used to detect the contour of Mycobacterium TB.

Algorithm 1 To find out image clusters that contain Mycobacterium tuberculosis.
  Let $X = \{x_1, x_2, x_3,..., x_n\}$ be the set of data points, Hue values of the image, and $V = \{v_1, v_2, v_3, v_4\}$ be the set of centers.

Step 1   Select 'c' cluster centers with the Hue values of pink, blue, white, and violet. Say $c = 4$

Step 2   Calculate the fuzzy membership 'μ_{ij}' using

$$\mu_{ij} = 1/\sum_{k=1}^{c=4} (d_{ij}/d_{ik})^{(2/m-1)} \qquad (6)$$

where μ_{ij} is the element of U-matrix and d_{ij} is the element of distance matrix, where d_{ij} is the Euclidian distance between ith object and jth cluster center, and m is a constant.

Step 3   Compute the new fuzzy centers 'v_j' using

$$v_j = \left(\sum_{i=1}^{n} (\mu_{ij})^m x_i\right) / \left(\sum_{i=1}^{n} (\mu_{ij})^m\right), \quad \forall j = 1, 2, \ldots c \qquad (7)$$

Step 4   Repeat Steps 2 and 3 until the minimum 'J' value is achieved or

$$\|U(k+1) - U(k)\| < \beta.$$

where

k is the iteration step.
β is the termination criterion between [0, 1].
$U = (\mu_{ij})n \times c$ is the fuzzy membership matrix.
J is the objective function.

Step 5   Find out all points which belong to the 'pink' cluster. Pink cluster contains the pixels of the image that contains the images of Mycobacterium TB bacteria. Figure 6 shows the output.

Step 6   Stop.

Fig. 6 Images of Mycobacterium tuberculosis bacteria after applying algorithm 1 upon Fig. 5

Fig. 7 The mask after opening

2.4.1 Remove the Noise Contours

The output image from Algorithm 1 contains some dots; they are basically noise. This small subcontours are removed by closing [14, 17], which is dilation followed by erosion. Erosion is able to remove unnecessary contours, which are small in size. The dilation of A by B, denoted by $A \oplus B$, is defined as the set operation

$$A \oplus B = \left\{ z | (\hat{B})_z \cap A \neq \phi \right\} \tag{8}$$

where $\emptyset$ is the empty set and B is the structuring element. In other words, the dilation of A by B is the set consisting of all the structuring element origin locations where the reflected and translated B overlaps at least one element of A. The erosion of A by B, denoted by $A \Theta B$, is defined as

$$A \Theta B = \left\{ z | (B)_z \cap A^c = \phi \right\} \tag{9}$$

Here, erosion of A by B is the set of all structuring element origin locations where no part of B overlaps the background of A. Figure 7 shows the cleared mask, and Fig. 8 shows the image where TB bacteria are visible.

2.5 Detection and Classification

The final step is detection and classification. If the number of bacteria present in the image is greater than 15, then severity is very high, and if it is less than 3, then severity is less, but affected by TB. The severity is determined by taking the ratio of the number of pixels used to show the Mycobacterium TB bacteria to the number of pixels present in the image.

Fig. 8 The clean images of Mycobacterium tuberculosis bacteria

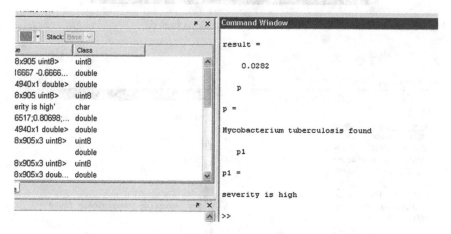

Fig. 9 The decision from the system for the input image shown in Fig. 1

Fig. 10 The new input image which is fed to the system

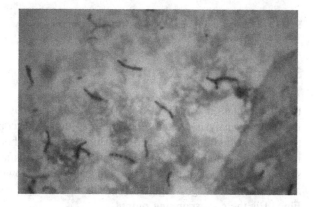

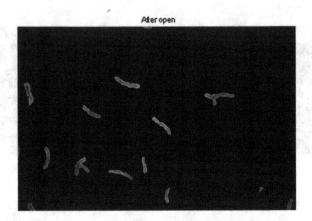

Fig. 11 The clean images of Mycobacterium tuberculosis bacteria

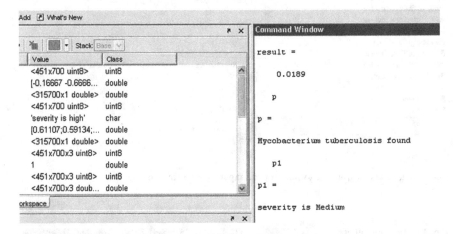

Fig. 12 The decision from the system for the input image shown in Fig. 10

3 Simulation and Results

For the simulation, MATLAB 7.1 [18] is used. The image shown in Fig. 1 is fed as input. After analysis, system finds the presence of Mycobacterium TB bacteria in the sample slide and the severity is high. The snapshot of the output is shown in Fig. 9. For another test image, the result is also satisfactory. The input image is shown in Fig. 10, and the corresponding output is shown in Fig. 12. Figure 11 is an intermediate stage. Figure 13 is an input image with no TB bacteria, and Fig. 14 is the corresponding output.

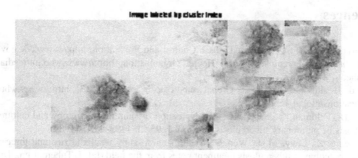

Fig. 13 The input image with no TB bacteria

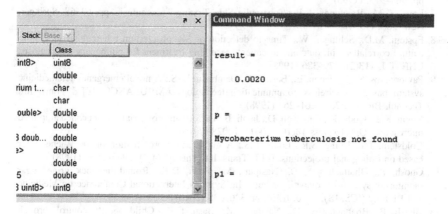

Fig. 14 The decision from the system for the input image shown in Fig. 13

4 Conclusion

In this paper, we have discussed a system for diagnosis of TB disease at any place and also through telemedicine system remotely. This system is cheaper than other test kits existing as present state of the art. This system does not require any special technical skill. So, this can be used by the people of remote places with very basic level of education. It may reduce the probability of wrong treatment which happens due to non-availability of diagnosis systems in remote and economically backward areas.

Acknowledgments Authors are thankful to Dr. Bera from K.C. Roy Tuberculosis Hospital, Kolkata, and Dr Soumendu Datta for providing pathological data with relevant knowledge, and the 'DST-GOI funded PURSE programme,' at Computer Science & Engineering Department, Jadavpur University, for providing infrastructure facilities during the progress of the work. Finally, we would like to thank Ms. Sweta Tiwari, Ms. Akanksha Ojha, Mr. Amit Roy Chowdury, and Ms. Aayatri Bhoumick; they helped a lot to collect samples.

References

1. Tuberculosis (TB).: Centers for Disease Control and Prevention, http://www.cdc.gov/tb/
2. Fact sheets.: Tuberculosis, World Health Organization, http://www.who.int/mediacentre/factsheets/fs104/en/
3. World Health Organization.: Global tuberculosis report 2013, http://www.who.int/tb/publications/global_report/en/
4. Levy, H., Feldman, C., Sacho, H.: A reevaluation of sputum microscopy and culture in the diagnosis of pulmonary tuberculosis. CHEST J. **95**(6), 1193–1197 (1989)
5. Feng-zeng, Z., Levy, M.H., Wen, S.: Sputum microscopy results at two and three months predict outcome of tuberculosis treatment notes from the field. Int. J. Tuberc. Lung Dis. **1**(6), 570–572 (1997)
6. Huebner, R.E., Maybelle F.S., Bass, J.B.: The tuberculin skin test. Clin. Infect. Dis., pp. 968–975. (1993)
7. Lordi, G.M., Reichman, L.B.: Tuberculin Skin Testing Tuberculosis, pp. 63–68. Springer, New York (1994)
8. Epstein, M.D., Schluger, W.: Time to detection of Mycobacterium tuberculosis in sputum culture correlates with outcome in patients receiving treatment for pulmonary tuberculosis. CHEST J. **113**(2), 379–386 (1998)
9. Pavlopoulos, S., Kyriacou, E., Berler, A., Dembeyiotis, S.: A novel emergency telemedicine system based on wireless communication technology-AMBULANCE. IEEE Trans. Inf. Technol. Biomed. **2**(4), 261–267 (1998)
10. Kevin, K.T., Goda, K., Capewell, D., Jalali, B.: Performance of serial time-encoded amplified microscope. Opt. Express **18**(10), 10016–10028 (2010)
11. Goldstein, J.S., Irving, S.R., Louis, L.S.: A multistage representation of the Wiener filter based on orthogonal projections. IEEE Trans. Inf. Theory **44**(7), 2943–2959 (1998)
12. Ghosh, P., Bhattacharjee, D., Nasipuri, M., Basu, D.K.: Round-the-clock urine sugar monitoring system for diabetic patients. In: Systems International Conference on Medicine and Biology (ICSMB), 2010. IEEE, pp 326–330 (2010)
13. Ghosh, P., Bhattacharjee, D., Nasipuri, M., Basu, D.K.: Child obesity control through computer game. Biomed. Eng. Res. **2**(2), 88–95 (2013)
14. Gonzalez, R.C., Richard, R.E.: Woods, digital image processing. 2nd edn. Prentice Hall Press, Upper Saddle River ISBN 0-201-18075-8 (2002)
15. Pedrycz, W., Waletzky, J.: Fuzzy clustering with partial supervision. IEEE Trans. Syst. Man Cybern. B Cybern. **27**(5), 787–795 (1997)
16. Pedrycz, W.: Algorithms of fuzzy clustering with partial supervision. Patt. Recog. Lett. Elsevier **2**(1), 13–20 (1985)
17. Ghosh, P., Bhattacharjee, D., Nasipuri, M., Basu, D.K.: Medical Aid for Automatic Detection of Malaria Computer Information Systems-Analysis and Technologies, pp. 170–178. Springer, Berlin (2011)
18. MATLAB User's Guide.: The mathworks. Inc., Natick, MA 5 (1998). http://www.mathworks.com

Automated Vertebral Segmentation from CT Images for Computation of Lumbolumbar Angle

Raka Kundu, Amlan Chakrabarti and Prasanna Lenka

Abstract Evaluation of lumbolumbar (LL) angle of the vertebral column from medical images is a common measure in case of patients suffering from lower back pain (LBP). There are five lumbar vertebrae (L1–L5). The angle formed between the slope of L1 and slope of L5 is the LL angle. This paper presents a computational technique for automated measurement of the LL angle from computerized tomography (CT) images using digital image processing techniques. The computation of LL angle comprises of four major steps: (i) image denoising (ii) segmentation of vertebral column (iii) morphological operations, and (iv) statistical estimation. To the best of our knowledge, the proposed technique for automated LL angle from CT images is the first of its kind. The segmentation based on s-t cut gives a promising result. We have performed our experiments over 30 CT images and have achieved satisfactory results in automated detection with an accuracy of 80 %.

Keywords Image processing · Lumbolumbar angle · Lumbar spine · Segmentation · s-t cut · Computer-aided diagnosis

1 Introduction

Lumbar vertebrae of human vertebral column carry major weight of the body. Lower back pain (LBP) [1] is a terminology that is commonly heard in orthopedics. It affects about 40 % [1] of people at some stage of their lives. Therefore, lumbolumbar (LL) angle [2] is a common measure that is followed to realize the

R. Kundu (✉) · A. Chakrabarti
A. K. Choudhury School of Information Technology, University of Calcutta,
Kolkata 700009, India
e-mail: kundu.raka@gmail.com

P. Lenka
R&D, National Institute for the Orthopaedically Handicapped, Kolkata 700090, India

© Springer India 2015
R. Chaki et al. (eds.), *Applied Computation and Security Systems*, Advances in Intelligent Systems and Computing 304, DOI 10.1007/978-81-322-1985-9_6

abnormality of lumbar region. It also helps in monitoring pre- and post-treatment condition of the patient. CT [3] and magnetic resonance imaging (MRI) [3] are recommended by doctors for diagnosis and for having a better realization about the disease.

Computer-aided diagnosis (CAD) makes the treatment procedure much fast, easy, and cost-effective. Many research works are intended in developing a reliable and CAD for vertebral abnormalities using CT and MR images. The edges of the vertebrae are quite prominent in CT and MR images, and they have a much better contrast. But still there lies a challenge for CT and MRI in totally automating the diagnosis technique without any user intervention. The pixel intensities of a particular region may be dissimilar; again pixel intensities of different regions may be quite similar. To make the computer-aided computation reliable, it is very essential to have a robust segmentation technique.

There are a good number of algorithms that have been proposed for vertebral segmentation in recent years. In [4], the authors proposed a method that automatically segments the spinal cord and canal from 3D CT images. For segmenting the spinal canal, an extended region-growing technique is suggested whereas for spinal cord segmentation active contours are applied. The algorithm needs one seed point in the central location of the spine where the seed point selection is for determination of starting slice and for providing the positional hint for segmentation. A vertebral segmentation technique from 2D CT images was proposed by Graf et al. [5]. The method possesses five steps: noise removal, region extraction and weighting, candidate generation, candidate selection, and dynamic refinement algorithm. In [6], the authors presented a research work on segmentation of vertebra from MR images based on multiple feature boundary classification and mesh inflation. But this segmentation technique was semi-automated, and a start point was needed for vertebral initialization. In recent years, graph cut [7] segmentation has gained popularity for its reliable results. Many research works can be found based on graph cut segmentation for evaluation of different diseases. In 2008, a graph cut approach for fully automated brain tumor segmentation in 3-D MRI was proposed by Wels et al. [8]. Cui [9] proposed a work on fully automatic segmentation of white matter lesions from MRI. Ababneh et al. [10] proposed a work based on graph cut from knee bones MRI for arthritis research. It is a content-based system that automatically segments knee bones and does not require user interaction. Evaluation of spine deformity in orthopedics carries a good importance in medical treatment. Research works are going on for computerized diagnosis of various spine deformities from medical images. Bhole et al. [11] proposed a technique for automatic detection of lumbar vertebrae and disk structure from MR images. Ghosh et al. [12] proposed an automated lumbar vertebral segmentation from clinical CT image that helps in wedge compression fracture diagnosis. The technique of computerized diagnosis was based on a collection of image processing techniques. Egger et al. [13] proposed a local template-based s-t cut segmentation. This research efficiently extracts the vertebrae from the MRI.

But, the major drawback of the technique lies in placing landmark (by user) on every vertebra for locating the individual template positions. This makes the total process time-consuming and laborious. The intensities that resemble the foreground and background are selected manually. In our research work, we have improved this template-based segmentation technique.

In this paper, we focus on CT images and have performed an automated computer-aided LL angle computation where the segmentation of the computer-aided technique is based on s-t cut. Our specific contributions are as follows:

- For segmentation, instead of considering all pixels of the image we have considered a single template whose dimension is similar to the image. The template distributes sampled nodes whose number is much less than the number of pixels of the image.
- The selection of source for s-t cut is done adaptively. The source intensity is computed from the histogram of the input image.
- The total procedure of LL angle evaluation is performed without any user intervention, and the research work aims in developing a totally automated CAD with high reliability.

The remaining sections of the paper are organized as follows. Section 2 describes the image processing techniques used in our proposed scheme of LL evaluation. Section 3 includes results and experimental data. Conclusion and future scope of research work are discussed in Sect. 4.

2 Proposed Method

2.1 Image Denoising

The medical CT images are sensitive to noise, which results in false intensity value. The contamination of image by noise may disturb the post-processing steps. So, a good denoising technique is necessary to make image noise free. State-of-the-art technique for image denoising is bilateral filtering [14]. It smoothes image as well as preserve edges. The performance of the filter is based on two kernels, the domain and the range kernel. The domain kernel is meant for noise removal in the homogeneous region, and the range kernel is designed for cleaning noise in discontinuous regions of the image. A mask of size 3×3 was selected for the denoising technique. A standard deviation of 3 and 0.1 was considered for the domain and range kernel of this preprocessing image enhancement. The segmentation is performed on this enhanced image. Denoised images after bilateral filtering are shown in Fig. 4a, c, and e.

2.2 Vertebral Segmentation

2.2.1 Graph Cut

Segmentation is partitioning the region of interest. We present a segmentation based on graph cut [7] that separates the vertebral column from the lateral view CT image. A template is considered that possess 7,200 points, which are sampled on the whole image. The CT image is rectangular in shape, and normally the height of the image is almost twice the width. It is observed that considering the row number as 120 and column number as 60 for the template gives a good segmentation result. This helps in maintaining an equidistant position between the nodes of the graph. The consideration of template for segmentation instead of all pixels of image allows reduction of computational complexity from N^2 to T^2 where $N \times N$ represents the dimension of the image and T^2 is the number of nodes in the template. Here, we have considered $T^2 = 7,200$, which is very small in comparison with number of pixels in an image.

The sampled points are the nodes $n \in V$ of the graph. The graph $G(V, E)$ is composed of set of nodes V and set of directed edges E. Every n of the graph possess grayscale intensity in the range [0 1]. The source s and sink t are the two virtual nodes of G. The basic principle of the proposed segmentation depends on intensities of s and t. There are E that connect the s and n of the graph and there are E between n and t. Intensity of s and t resembles the intensity of foreground (vertebral column) and background, respectively. Every E of the graph is assigned a positive weight w. n that has intensity value more similar to s will have more weight for the E that connects between s and n where as in that case n will have less weight for the E that connects the same n to the t. The reverse principle is applicable when the intensity value of n is more similar to the intensity of t. In our segmentation technique, the mathematical expression of weights (w) for s and t are as follows:

$$w_s = \exp\left(-\frac{d(s,n)}{h}\right) \tag{1}$$

$$w_t = \exp\left(-\frac{d(t,n)}{h}\right) \tag{2}$$

w_s and w_t are the weights for s and t, $d(s, n)$ is the intensity difference between s and n, and $d(t, n)$ is the intensity difference between t and n. h is the control parameter of the weight.

The s-t cut is grouping the n of the graph (template) into two subsets S and T such that n similar to s lies in S and the n similar to t lies in T. Figure 1 shows a simple example of the theory.

Fig. 1 **a** Template for *s-t* cut
segmentation, where the color
indicates the *gray color*
intensity that the nodes
possess. **b** Nodes having
intensity similar to *s* have
more weighted edge for *s* and
vice versa. **c** Nodes of the
template selected in the group
of foreground are indicated in
white and nodes selected in
the group of background are
indicated in *black*. **d** Figure
represents only the border
nodes of the foreground-
segmented regions. The
border nodes are marked
white

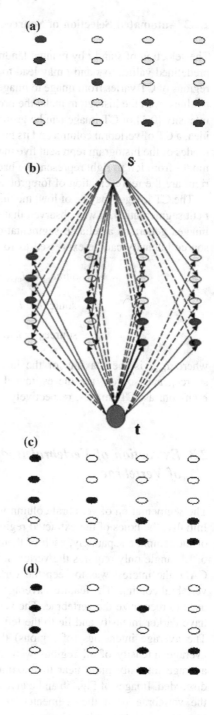

2.2.2 Automated Selection of Source

The selection of s and t by manual landmarks is a tedious process. Even assigning predefined values to s and t may lead to erroneous results because the intensity of regions of CT varies from image to image. We have observed that selection of s and t values from the histogram make the procedure much adaptive. The wide range of intensity [0 1] of CT image can be grouped into five major subgroups. To have an idea, a CT of vertebral column and its histogram are shown in Fig. 2, where the five modes of the histogram represent five major subgroups of the image. The first three modes from left to right represent the background, and the remaining modes on the right are the representation of foreground (vertebral column).

The CT images may be of low, medium, and high contrast. For the automated s-t cut segmentation, it was observed that sink value of 0.15 for all types of contrast images generates a reliable segmentation. Whereas the formula for selection of source intensity can be expressed as follows:

$$source_m = I_{max} + (I_{max} - t) \times 2.3 \tag{3}$$

$$source_l = I_{max} + (I_{max} - t) \times 2 \tag{4}$$

$$source_h = I_{max} + (I_{max} - t) \times 4 \tag{5}$$

where I_{max} is the maxima of the histogram above intensity 0.1 and $source_m$, $source_l$, and $source_h$ are the estimated source intensity for medium-, low-, and high-contrast CT images, respectively.

2.3 Extraction of Vertebral Body and End Plates of Vertebrae

The segmentation of vertebral column was followed by morphological operations. Initially, the holes of the extracted regions were filled. And to identify the regions of the template separately, each of them were labeled uniquely. The computation of LL angle only requires the vertebral body end plate. So in the next step of the CAD, the interest was to keep the vertebral body and discard other parts of the vertebral column. The features, average intensity, and location of the regions were used to recognize the vertebrae. The vertebrae of the segmented vertebral column have darker intensity and lie to the left than other segmented foreground regions. The average intensities (of regions) those were more similar to the minimum average intensity of all regions were considered as vertebrae and regions with average intensity more near to maximum average intensity of all regions were discarded. Images of Fig. 5b and c give clear idea of the section. After selection of the vertebrae from the segmented regions, morphological operations such as removing isolated nodes, removing small regions, extraction of region boundary,

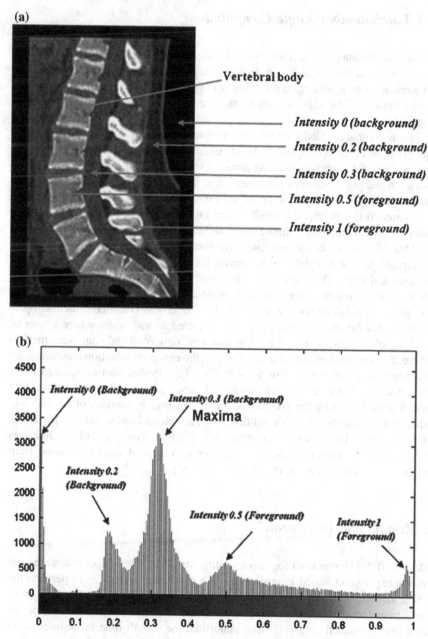

Fig. 2 **a** CT image of vertebral column and **b** Corresponding histogram

and removal of vertical lines were performed to extract end plates of the vertebrae. All these processing were done with the nodes of the template, where the template was superimposed on the image.

2.4 Lumbolumbar Angle Computation

In manual measure of LL angle, pencil and scale are used to draw lines on the printed medical images. A line is drawn through the end plate corner points of the L1 vertebra, and similarly another line is drawn through the end plate corners of the L5 vertebra. The angle formed by the intersection of these two lines is the LL angle [2].

In our computer-aided process, the extracted end plates were relabeled so that end plates of every vertebra could be identified. As there are five lumbar vertebrae and every vertebra possess two end plates, it was observed that selection of second and tenth end plate (bottom to up direction of the template) helped us in getting the plates that represented the slope of extreme vertebrae (L1, L5) of the lumbar curvature. At this point, an intensity value of 1 was assigned to the nodes of one end plate and an intensity value of 2 was assigned to the nodes of second end plate. And the other nodes of the template were assigned value 0. Next, a scanning was performed on the template to determine the end nodes of each of the finally selected end plate. The slope of line obtained by joining the end nodes of an end plate can be considered similar to line obtained by joining the corner points of the end plates. A neighborhood window of size 3×3 was considered for this operation. Nodes for the first end plate were selected as end nodes where a sum of intensity of the window was 2 and end nodes of the second end plate were marked where the sum resulted as 4. Figure 3 shows the end node selection conditions for end plates one and two. It can be seen from Fig. 3b that sum of values of any condition is 2. If a node satisfies any one of these condition, it is end node of first line. Figure 3c shows the conditions for considering end nodes of second line where sum of values of every condition is 4. The selected end nodes of a particular end plate were joined with each other to get a line. The angle of the line was computed with respect to the horizontal axis. A sum of angles obtained from superior and inferior end plate gave the LL angle.

3 Results and Discussion

MATLAB 2011b was used for implementing the CAD. The images that were used for the experiment varied in resolution. As a result, the time taken to perform the image denoising varied from one CT image to other. The post-processing steps were based on fixed number of template nodes. So, the computational time for later steps remained constant. After denoising, the overall time taken to perform the template-based segmentation and computation of LL angle from nodes was 7 s. A machine specification of Intel Core2 Duo CPU, 3 GHz, 3 GB RAM, Windows XP Professional Version 2002, and Service Pack 3 was used for this work. The time required to process the whole CAD on CT was about 14–25 s.

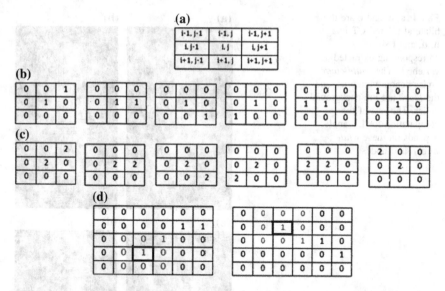

Fig. 3 **a** Shows the 3 × 3 neighborhood. **b** Illustrates conditions for regarding a node as end node of first line. **c** Shows the conditions of end node for second line. **d** Shows example of identification of a node as end node when it matches any one of the above conditions

The vertebral body segmentation results from different patient's CT image are shown in Fig. 4a, c, and e show the denoised images, and b, d, and f show the corresponding extracted vertebrae. Due to similar average intensity of regions (explained in Sect. 2.3), there remains a chance that sometimes some other parts of the vertebral column get extracted with vertebrae. Such an example is shown in Fig. 4d. As the main objects of interest are horizontal end plate lines from the square vertebrae, it is observed that post-morphological operations can easily remove these unwanted regions.

Figure 5 shows the results of main steps of our proposed technique. The final image of the figure shows the identified slope of extreme vertebrae (L1, L5) for the lumbar curvature.

Next, we have performed an experiment to judge the segmentation accuracy. A direct comparison is done between the ground truth developed by manual landmarks and the segmented vertebra obtained from our proposed *s-t* cut. The comparison is shown in Fig. 6. The white-colored nodes of Fig. 6b are the ground truth of vertebra boundary, and Fig. 6c is our segmentation result on the ground truth image, Fig. 6b. Figure 6c shows that the ground truth of vertebra boundary directly matches with the boundary of the automated segmented vertebra (white).

The LL angles (in degree) computed from our automated CAD are shown in Table 1. We have performed the experiment of computerized measurement over low-, medium-, and high-contrast CT images. In every case, manual measurement [2] of LL angle over printed CT image was recorded. For this, lines were drawn on the end plate of the upper lumbar vertebra and on the end plate of the lower lumbar

Fig. 4 **a**, **c**, and **e** are the bilateral filtered CT images. **b**, **d**, and **f** show corresponding extracted vertebrae. The *white-colored* nodes denote vertebrae, and the *black* are the background nodes. The rest of the computation for LL angle depends on these *white* foreground nodes

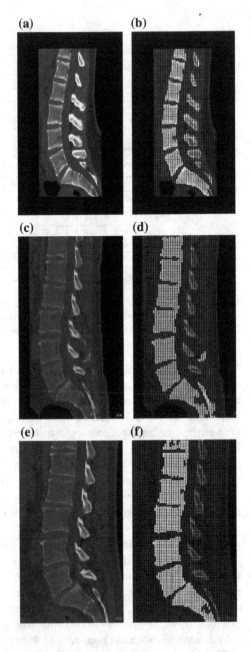

vertebra. The interior angle formed by the intersection of these two lines is the LL angle. Manual LL angle measurement may lead to some variability if the lines that are running over the end plate of the vertebra are not drawn exactly through the extreme corners of the end plate. An intra-observer CAD was recorded where two trials were taken by the user for every CT image. Considering the manually

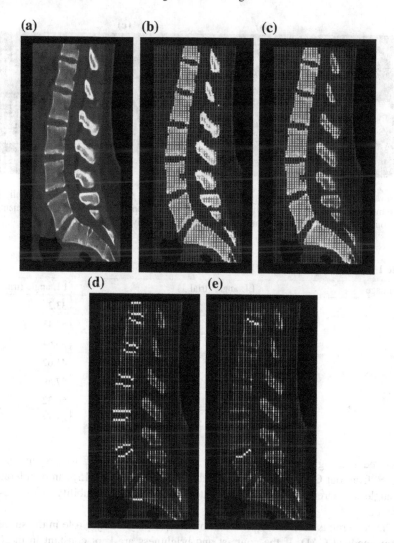

Fig. 5 Steps of our technique is as follows : **a** denoised CT image and **b** shows the automated segmented vertebral column based on *s-t* cut. **c** Shows the selected vertebral body (vertebrae) from (**b**). The *white nodes* of **d** show the extracted end plate lines of the vertebrae from (**c**), and **e** shows the identified slopes of the superior and inferior lumbar vertebrae. The LL angle is evaluated by computing the angle formed by intersection of these two lines

(a)   **(b)**   **(c)**

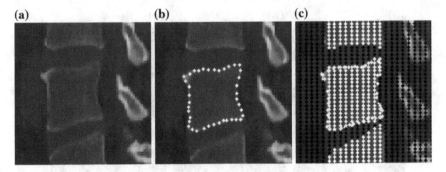

Fig. 6 **a** Shows the input image and **b** is ground truth. *White points* were marked manually on the boundary of middle vertebra of the image **a**. **c** Shows the superimposed automated segmented result on the ground truth where both the boundaries efficiently match each other

Table 1 LL angle value from CAD

CT image	LL angle (trial 1)	LL angle (trial 2)
1	33.1	33.2
2	53.31	57.43
3	36.64	36.64
4	46.55	41.62
5	39.21	37.59
6	51.87	50.32
7	41.26	41.15

measured LL angle as standard value, a variability of $\pm 0.1°$ to $\pm 5°$ has been noticed from our CAD technique. In medical an error of $\pm 5°$ [15] can be tolerable for angle measurement. This demonstrates the clinical acceptability of our CAD technique.

The constraints such as contrast and brightness play a major role in the success of any medical CAD. If the contrast and brightness are kept constant in the CT imaging, it will further help in reducing the variability of our CAD to a much greater extent.

4 Conclusion

In this paper, we have presented an automated LL angle evaluation process from clinical CT image. Initially, the source intensity is computed from histogram, and segmentation is performed using template-based *s-t* cut. A proper selection of *s* and *t* gives a very reliable segmentation result. The automated CAD of LL angle from CT image reduces user intervention. This research work is a good initiative for automated clinical application. Digital measurement also eases up the processing as it can be entirely done on the system, which captures the source image. In future, we will try to reduce the variability of this computer-aided evaluation and also we will try to extend our work of automated LL angle computation to digital X-ray image cases.

References

1. Low back pain, from Wikipedia. http://en.wikipedia.org/wiki/Low_back_pain
2. Damasceno, L.H.F., Catarin, S.R.G., Campos, A.D., Defino, H.L.A.: Lumbar lordosis: a study of angle values and of vertebral bodies and intervertebral discs role. Acta. Ortop. Bras. **14**(4), 193–198 (2006)
3. Wilmink, J.T.: Lumbar Spinal Imaging in Radicular Pain and Related Conditions, pp. 9–30. Springer, Berlin (2010)
4. Nyul, L.G., Kanyo, J., Mate, E., Makay, G., Balogh, E., Fidrich, M., Kuba, A.: Method for automatically segmenting the spinal cord and canal from 3D CT images. In: 11th International Conference on Computer Analysis of Images and Patterns, pp. 456–463. Springer, Berlin, Heidelberg (2005)
5. Grafa, F., Greila, R., Kriegela, H., Schuberta, M., Cavallarob, A.: Enhanced detection of the vertebrae in 2D CT-images. In: Medical Imaging, SPIE digital library (2012)
6. Zukic, D., Vlasak, A., Dukatz, T., Egger, J., Horinek, D., Nimsky, C., Kolb, A.: Segmentation of Vertebral Bodies in MR Images, Vision, Modeling, and Visualization, pp. 135–142. The Eurographics Association, Germany (2012)
7. Boykov, Y., Veksle, O.: The Grid: Handbook of Mathematical Models in Computer Vision, pp. 79–96. Springer, US (2006)
8. Wels, M., Carneiro, G., Aplas, A., Huber, M., Hornegger J., Comaniciu, D.: A discriminative model-constrained graph cuts approach to fully automated pediatric brain tumor segmentation in 3-D MRI. Med. Image Comput. Comput. Assist. Interv. 67–75 (2008)
9. Cui, S.: Fully Automatic Segmentation of White Matter Lesions from Multispectral Magnetic Resonance Imaging Data. Department of Information Technology, Institutionen for informationsteknologi
10. Ababneh, S.Y., Prescott, J.W., Gurcan, M.N.: Automatic graph-cut based segmentation of bones from knee magnetic resonance images for osteoarthritis research. J. Med. Image Anal. **15**, 438–448 (2011)
11. Bhole, C., Kompalli, S., Chaudhary, V.: Context sensitive labeling of spinal structure in MR images. In: Medical Imaging, SPIE digital library (2009)
12. Subarna, G., Raja, A., Chaudhary, V., Dhillon, G.: Automatic lumbar vertebra segmentation from clinical CT for wedge compression fracture diagnosis. In: Proceedings of the SPIE Medical Imaging (2011)
13. Egger, J., Kapur, T., Dukatz, T., Kolodziej, M., Zukic, D., Freisleben, B., Nimsky, C.: Square-cut: a segmentation algorithm on the basis of a rectangle shape. J. PLoS One. **7** (2012)
14. Tomasi, C., Manduchi, R.: Bilateral filtering for gray and color images. In: Proceedings of Sixth International Conference on Computer Vision, ICCV, pp. 839–846 (1998)
15. Bruton, A., Conway, J.H., Holgate, S.T.: Reliability: what is it, and how is it measured. Physiotherapy **86**, 94–99 (2000)
16. Sagittal lumbar spine, from fairfaxradiology. http://www.fairfaxradiology.com/lowdose/CS_MS.php
17. From naplesxray. http://www.naplesxray.com/ct-head-neck-spine-extremities.html

An Approach for Micro-Tomography Obtained Medical Image Segmentation

Mateusz Buczkowski, Khalid Saeed, Jacek Tarasiuk, Sebastian Wroński and Joanna Kosior

Abstract This paper presents a method for segmentation of micro-tomography images. Proper segmentation of those images is necessary to create visualization of these structures. The introduced algorithm concerns finding the proper way of image filtering before the use of Canny-Deriche edge detection to obtain the best possible segmentation.

Keywords Bilateral filter · Image processing · Image segmentation · Edge detection · Median filter · Gauss filter · Canny-Deriche edge detector · Smoothing · Micro-tomography · Porous materials

1 Introduction

Computed tomography (CT) is a powerful, modern technique for nondestructive internal structures imaging of any object. Classical CT was invented in 1967 by Sir Godfrey Hounsfield. The method is continuously developing giving more precise

M. Buczkowski (✉) · K. Saeed · J. Tarasiuk · S. Wroński · J. Kosior
Faculty of Physics and Applied Computer Science, AGH University of Science and Technology, Krakow, Poland
e-mail: mateusz.buczkowski1@gmail.com

K. Saeed
e-mail: saeed@agh.edu.pl

J. Tarasiuk
e-mail: tarasiuk@agh.edu.pl

S. Wroński
e-mail: wronski@agh.edu.pl

J. Kosior
e-mail: joanna.kosior0@gmail.com

K. Saeed
Faculty of Computer Science, Bialystok University of Technology, Bialystok, Poland

© Springer India 2015
R. Chaki et al. (eds.), *Applied Computation and Security Systems*, Advances in Intelligent Systems and Computing 304, DOI 10.1007/978-81-322-1985-9_7

and accurate pictures in shorter time and using smaller radiation dose. Till the end of XX century, the main areas of CT applications were bioimaging in medicine and industry inspection. For 15 years, the new micro-computed tomography (μCT) devices with much higher resolution have become more and more popular. Today, the best μCT equipment can reach the resolution better than 1 μm. Because of their high-research potential, the μCT wins last year many scientific areas such as biology, geology, material science, and so on. The general idea of μCT measurements is very simple. X-ray tube is generating radiation penetrating the sample. The 2D detector on the opposite side of the sample is registering attenuation of the X-ray intensity. The registered picture is called 'projection'. The value of attenuation in each pixel of the detector depends on the materials property across the single ray. The sample is rotating, and few hundred or thousand projections are registered. After measurement, the computer program is used to reconstruct 3D object from set of 2D projections. Exist several methods to do that. One of the most popular is filtered back projection method based on Radon transform theorem [1]. The reconstructed object is represented as a three-dimensional matrix of voxels. The voxels may be represented by 8-, 16-, or 32-bit values. For visualization of 3D objects, very often isosurface rendering or volume-rendering methods are used. But for precise analysis including dimension measurements or some feature extractions, the cross sections in selected direction are used. Depending on the research subject, different methods of analysis are applied.

The X-ray measurements presented in this paper were performed using 'nanotom 180 N' device produced by GE Sensing & Inspection Technologies phoenix|X-ray Gmbh. The machine is equipped with nanofocus X-ray tube with maximum 180 kV voltage. The tomograms were registered on Hamamatsu 2300 × 2300 pixel detector. The reconstruction of measured objects was done with the aid of proprietary GE software datosX ver. 2.1.0 with use of Feldkamp algorithm for cone beam X-ray CT [2]. The post-reconstruction data treatment was performed using VGStudio Max 2.1 [3] and free Fiji software [4].

In this publication, we will concentrate on a single, important problem in porous materials study. The typical porous material can be described as a two-phase composite, where one phase is a solid or condensed phase and the second one is a void or some gas or liquid phase. The problem is to separate on each cross section the solid and the void part of the material. In the case of high difference in X-ray linear attenuation factor for both phases (like porous titanium for example), the solution is very simple. The histogram for voxels values consists of two separated Gaussian-shaped peaks. Any threshold value between the peaks gives correct segmentation for solid and void phases. For some other materials usually minerals, the peaks on the histograms overlap in some parts (Fig. 1). Few median filters applied to the cross section leads to more separated peaks (Fig. 2). Now some binarization methods (like Otsu) may be used to obtain separated phases (Fig. 3).

The situation is little more complicated for less contrast (contrast for X-ray) materials such as trabecular bone (both human and animals). The bone is built from the compact/cortical bone (solid phase) and the marrow (liquid phase).

(a)

(b)

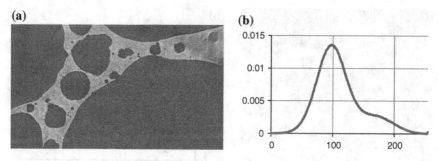

Fig. 1 Micro-tomography image of mineral (**a**) and its histogram (**b**)

(a)

(b)

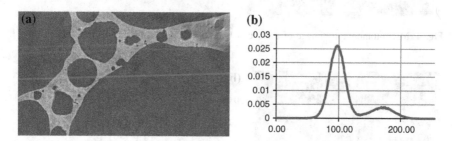

Fig. 2 Micro-tomography image of mineral with median filtering (**a**) and its histogram (**b**)

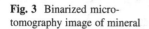

Fig. 3 Binarized micro-tomography image of mineral

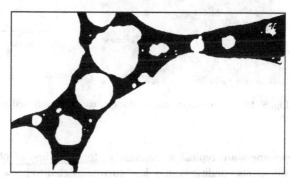

Typically on the cross sections, both phases are easily recognizable but on the histogram only one much broadened peak exists (Fig. 4).

Also in this case, few median filters (Fig. 5) and Otsu method application lead to clear separation of the phases.

The final 3D reconstruction is consistent, and the solid phase is well visible (Fig. 6). Some soft materials, almost transparent for X-ray radiation represent the most difficult case. The biological soft tissue, floral samples, or many lightweight

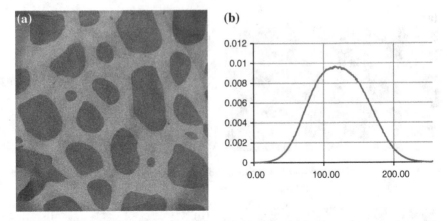

Fig. 4 Micro-tomography image of trabecular bone (**a**) and its histogram (**b**)

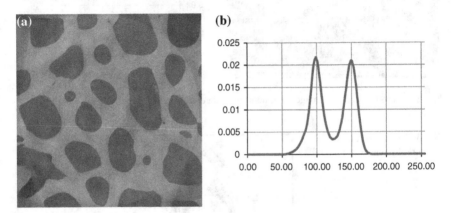

Fig. 5 Micro-tomography image of trabecular bone with median filtering (**a**) and its histogram (**b**)

polymers are typical representative for this class of objects. Also the bad measurements conditions may lead to weak signal-to-noise ratio.

Typical examples are shown below: the pumice measured in bad conditions (Fig. 7) and the polymer foam (Fig. 8).

The solid phase structures are visible on the pictures, but on the histogram only one peak exists. Furthermore, any typical filter applied to the image cannot separate the phases, because voxels with the same grey values exist inside and outside of the solid phase. Any threshold value will lead to high fragmentation of the solid phase or to appearing of the false solid phase inside the void phase.

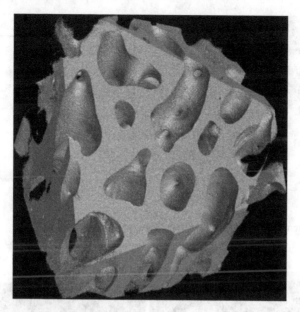

Fig. 6 3D reconstruction of trabecular bone

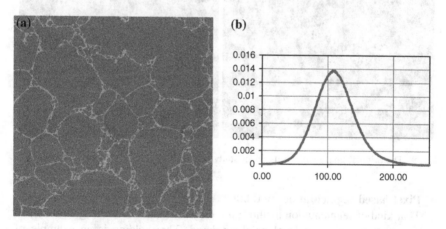

Fig. 7 Micro-tomography image of pumice measured in bad conditions (**a**) and its histogram (**b**)

2 State of the Art

There is no direct way for image segmentation [5]. The most common segmentation methods belong to one of the following methodologies [6]:

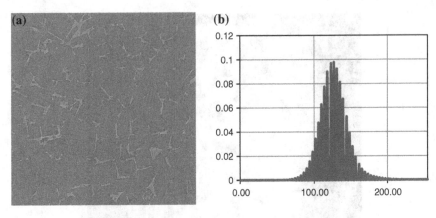

Fig. 8 Micro-tomography image of polymer foam (**a**) and its histogram (**b**)

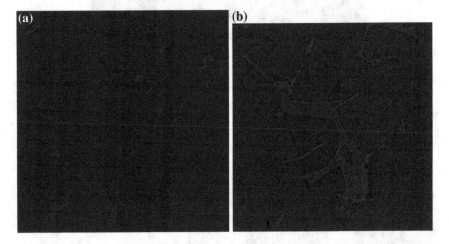

Fig. 9 The original image (**a**) and its under study part (**b**)

- **Pixel-based segmentation methods**
  That kind of segmentation methods use only the values of the single pixels and do not concern about local neighbourhood. Thresholding is an example of pixel-based method [7].
- **Region-based segmentation methods**
  That type of methods analyse the values in larger areas [8]. Neighbouring pixels are checked, and if they fulfil some conditions of similarity, they are added to form homogeneous region. Region-growing algorithm is an example of region-based method.

Fig. 10 Image after thresholding with one threshold

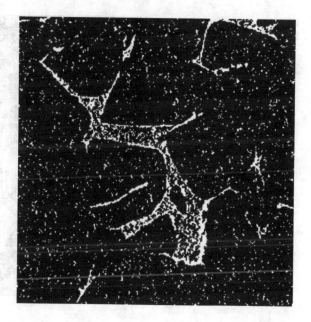

- **Edge-based segmentation methods**
  These methods use information about gradient to detect edges and then try to follow them. Canny-Deriche method is an example of edge-based method [9].
- **Model-based segmentation methods**
  That kind of methods are imitating some physical properties or processes and tries to find object contour by using information from image. Deformable models show model-based example [10, 11].

Deformable models are widely and successful applied for medical image segmentation. They give good results when studied images are corrupted by noise and artefacts. On such images, techniques as thresholding or edge detection may have difficulties with proper segmentation [10]. Various methods based on deformable models were worked out [11].

2.1 Some Existing Approaches Applied to Our Sample Image

Sample segmentation methods have been tested. Comparison of methods has been performed on image which is a part of original image (Fig. 9). Results shown in the next figures prove that classical methods alone fail in segmentation of images studied in this paper.

Simple thresholding fails because of similarities in intensities between pixels inside objects and background (Fig. 10).

Fig. 11 Image after use of
Sobel filter

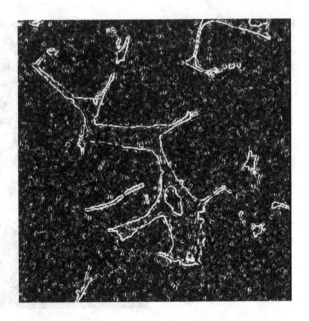

Sobel filter is sensitive to noise what in our case produces very large amount of
false-detected edges (Fig. 11).

Moreover, edges are not thinned. Canny-Deriche edge detector (Fig. 12) with
low-smoothing level produces significant amount of false-detected edges and
leaves some gaps in the edges.

Canny-Deriche edge detector [9, 12] with high-smoothing level produces small
amount of false-detected edges but creates large gaps in edges (Fig. 13). However,
authors' algorithm detects edges more optimally with small amount of false-
detected edges, and small gaps as will be seen at the end of Sect. 4.

3 Used Methods

3.1 Bilateral Filter

Bilateral filter is a smoothing technique which allows to remove textures, noise,
details but preserves large edges without blurring. Bilateral filter is a modified
version of Gaussian convolution. Gaussian filtering is weighted average of the
adjacent pixels intensities in the given neighbourhood where weights are
decreasing with increasing spatial distance from the central pixel as given in (1).

$$G[I]_p = \frac{1}{W_{pG}} \sum_{q \in S} G_\sigma(||p - q||)I_q, \tag{1}$$

Fig. 12 Image after use of Canny–Deriche edge detector with low-smoothing level

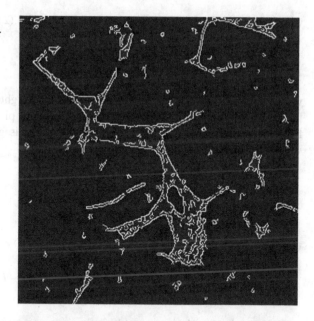

Fig. 13 Image after use of Canny–Deriche edge detector with high-smoothing level

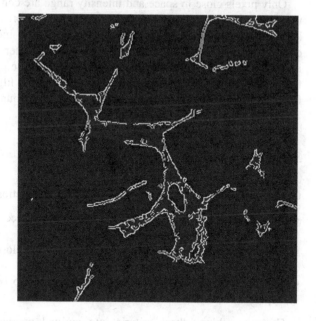

where $G_\sigma(x)$ is Gaussian convolution kernel given by (2).

$$G_\sigma(x) = \frac{1}{2\pi\sigma^2}\exp\left(-\frac{x^2}{2\sigma^2}\right). \tag{2}$$

where: S is the spatial domain, W_{pG} is sum of all weights, I is intensity of pixel, $\|p - q\|$ is the Euclidean distance between the central pixel p and another pixel q form the given neighbourhood. Neighbourhood size is given by σ. Main disadvantage of Gaussian filtering is edge blurring.

Bilateral filter is defined as in (3)

$$\mathrm{B}[I]_p = \frac{1}{W_{pB}}\sum_{q\in S}G_{\sigma_s}(\|p - q\|)G_{\sigma_r}(\|I_p - I_q\|)I_q, \tag{3}$$

where,

$$W_{pB} = \sum_{q\in S}G_{\sigma_s}(\|p - q\|)G_{\sigma_r}(\|I_p - I_q\|). \tag{4}$$

Only pixels close in space and intensity range are considered. Gaussian kernel in spatial domain (G_{σ_s}) decreases weights with increasing distance. Gaussian kernel in range domain (G_{σ_r}) decreases weights with increasing intensity differences. Combination of both kernels give bilateral filter capability of smoothing image and preserve edges at the same time [13]. These properties are very useful for smoothing μCT images studied in this paper. That filter is especially effective for images containing small objects like those under our consideration.

3.2 Canny-Deriche Edge Detector

Canny formulated the criteria for effective edge detection:

- Good detection—low probability of failing to detect existing edges and low probability of false detection of edges
- Good localization—detected edges should be as close as possible to the true edges
- One response to one edge—multiply responses to one real edge should not appear

Canny combined these criteria into optimal operator and found that it is approximately the first derivative of Gaussian [9, 12]—see (5)

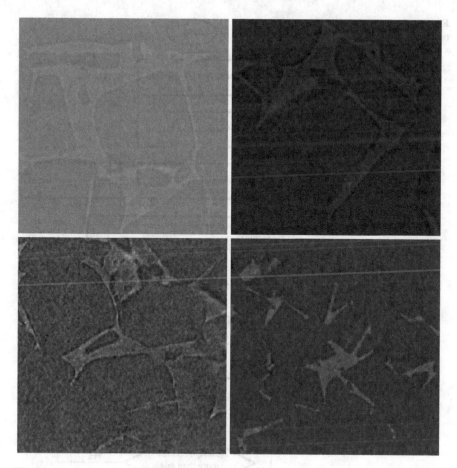

Fig. 14 Some examples of noise type, various images intensity, and similarities in pixels intensities inside objects of interest and background

$$f(x) = -\frac{x}{\sigma^2} e^{-\frac{x^2}{2}\sigma^2}. \tag{5}$$

In 1987, Deriche-modified Canny approach to implement optimal edge detector [9]. He presented optimal edge detector with impulse response given by:

$$f(x) = k \cdot e^{-\alpha \cdot |x|} \sin \omega x \tag{6}$$

and second version when ω tends to 0:

$$g(x) = k \cdot x e^{-\alpha \cdot |x|}. \tag{7}$$

Fig. 15 Flow chart of
proposed approach

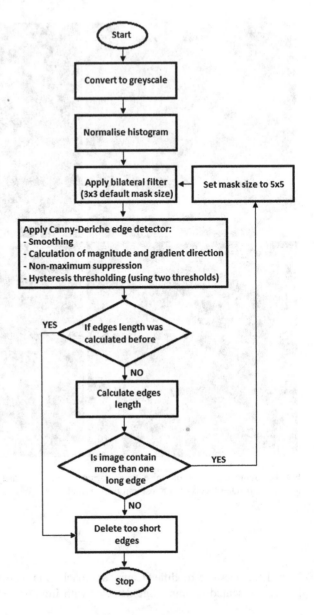

Performance of that approach is much better than the first derivative of a Gaussian presented by Canny. Calculation of magnitude and gradient direction are performed. Then, non-maximal suppression selects the single maximum point across the width of an edge. After that, step edges are thinned. Then, hysteresis thresholding is performed to get the final result. Hysteresis thresholding using two threshold divides pixels into three groups. Pixels below low threshold are

Fig. 16 Image after conversion to grey scale and histogram normalization

removed, and above high threshold are retained. Pixel with intensity between low and high threshold is retained only if connected to pixel above high threshold [9, 12].

4 The Proposed Methodology

Images which are analysed in this paper are quite complicated to segment. The objects size, image size, image intensity, and level of noise are various between each individual group of images. Some of them include artefacts resulting from the method used. The major difficulty is that some of the objects have intensities of pixels very similar to the background elements (Fig. 14). That inconvenience rules out use of the simple segmentation methods based on pixel intensity such as thresholding. This paper focuses on edge detection.

Figure 15 shows the flowchart of the suggested approach.

First histogram normalization is applied to the original image. Due to high noise level of images, efficient smoothing plays a key role. Our approach involves bilateral filter. After conversion to greyscale and histogram normalization (Fig. 16), bilateral filter is applied (Fig. 17). Bilateral filter uses 3×3 mask by default. Then, Canny-Deriche edge detector is applied. If the image contains more than one long edge (above threshold), then objects from this image are considered as big. Otherwise, too short edges are removed which leads to the final result. For images considered as including big objects, procedure starts over with the use of

Fig. 17 Image after use of
bilateral filter with 5 × 5
mask size

bilateral filter with 5 × 5 mask in spatial distance domain omitting calculation of
edges length.

The mask size is based on segmented objects size. Canny–Deriche algorithm [9,
12] is used to detect edges (Fig. 18).

Parameters are fixed. In the last step, small false-detected edges are removed
(Fig. 19). Algorithm removes only edges shorter than the chosen value. That the
value is based on object size.

5 Experimental Results and Interpretation

The algorithm produces very small amount of false edges detected, but those very
weak and hard to differ from background are not detected. Using bilateral filtering
prior to Canny–Deriche edge detector gives better results than Canny–Deriche
algorithm itself. Some objects are not fully segmented because of some gaps in
edges occurred after detection. Parameters of Canny–Deriche edge detector are
fixed similarly as bilateral filter intensity range. Bilateral filter spatial distance is
adjusted to object size. We hope that the approach proposed after some modifi-
cations could give fully segmented images. The basic feature of the presented
method is its ability to treat weak contrast images giving promising results of
image segmentation (Figs. 20, 21, 22, 23).

Fig. 18 Image after use of Canny–Deriche edge detector

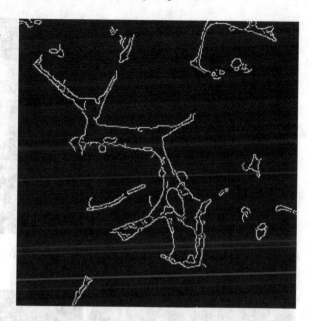

Fig. 19 Image after removal of too short edges

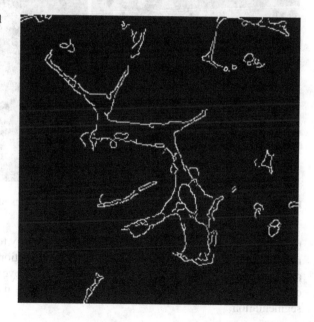

In some cases, it is impossible to get good-quality images. One of the samples is the biological soft tissue which is drying during measurements. To prevent the sample from being dehydrated, the measuring time is decreased. But short time for a single projection leads to a decrease in the signal-to-noise ratio. Newly proposed

Fig. 20 The original sample image (**a**) and segmentation results of the whole image (**b**)

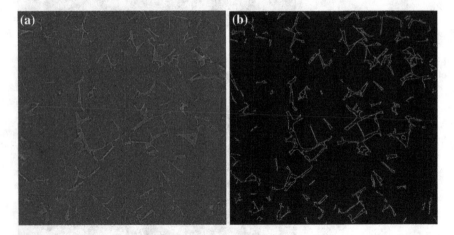

Fig. 21 The original sample image (**a**) and segmentation results of the whole image (**b**)

method for cross-sectional processing gives a possibility to significantly reduce the measurement time with similar quality of phase detection as longer measurement and classical workflow based on median filters. Moreover, the proposed approach can help to determine initial contours for deformable models methods of image segmentation.

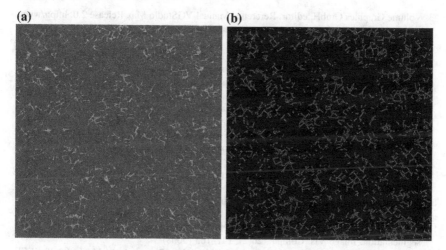

Fig. 22 The original sample image (a) and segmentation results of the whole image (b)

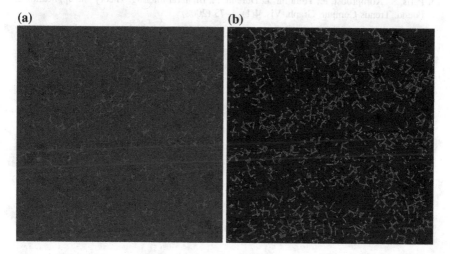

Fig. 23 The original sample image (a) and segmentation results of the whole image (b)

Acknowledgments The research was partially supported by grant no. WFiIS 11.11.220.01 AGH University of Science and Technology in Krakow.

References

1. Radon, J.: Uber die Bestimmung von Funktionen durch ihre Integralwerte Langs Gewisser Mannigfaltigkeiten. Ber. Saechsische Akad. Wiss. **29**, 262 (1917)
2. Feldkamp, L.A., Davis, L.C., Kress, J.W.: Practical cone-beam algorithm. J. Opt. Soc. Am. **A6**, 612–619 (1984)

3. Volume Graphics GmbH, editor. Reference Manual VGStudio Max Release 2.0; http://www. volumegraphics.com/en/products/vgstudio-max/ 8.10.2013
4. http://fiji.sc/Fiji
5. Rogowska, J.: Overview and fundamentals of medical image segmentation handbook of medical imaging, pp. 69–85. Academic Press Inc, New York (2000)
6. Jahne, B.: Digital image processing: Concept, algorithms, and scientific applications. Springer, New York (1997)
7. Sezgin, M., Sankur, B.: Survey over image thresholding techniques and quantitative performance evaluation. J. Electron. Imaging 13(1), 146–165 (2004)
8. Saeed, K., AlBakoor, M.: Region growing based segmentation algorithm for typewritten, and handwritten text recognition. Appl. Soft Comput., Elsevier Sci. Publishers 9(2), 608–617 (2009)
9. Deriche, R.: Using Canny's criteria to derive a recursively implemented optimal edge detector. Int. J. Comput. Vis. 1(2), 167–187 (1987)
10. Chenyang, Xu, Pham, D.L., Prince, J.L.: Image segmentation using deformable models. Handb. med. imaging 2, 129–174 (2000)
11. He, L., et al.: A comparative study of deformable contour methods on medical image segmentation. Image Vis. Comput. 26(2), 141–163 (2008)
12. Canny, J.: A computational approach to edge detection. Pattern Anal. Mach. Intell., IEEE Trans. 6, 679–698 (1986)
13. Paris, S., Kornprobst, P., Tumblin, J., Durand, F.: Bilateral filtering: Theory and applications. Found. Trends Comput. Graph. Vis. 4(1), 1–73 (2008)

Flexible Cloud Architecture
for Healthcare Applications

Amit Kr Mandal, Anirban Sarkar and Nabendu Chaki

Abstract A healthcare cloud is used by healthcare service providers for storing, maintaining, and backing up personal health information along with structured management of the health data across multiple healthcare providers. On a daily basis, healthcare services deal with different kinds of digital information ranging from structured to unstructured. The widespread adoption of electronic health records have resulted in an improved patient health and safety as well as significant savings in healthcare costs. Moreover, deploying healthcare records over cloud environment will enable access of critical patient's information at any time and from anywhere. But many enterprises are facing a major research challenge due to the unavailability of suitable cloud architecture for design, development, and deployment of healthcare services. In this paper, a flexible architecture for SaaS-based healthcare services has been proposed specifically for healthcare applications, which is capable of semi-structured healthcare data management and storing compatible with Health Level Seven (HL7) standard (Hennessy et al. in A framework and ontology for mobile sensor platforms in home health management, 2013; Liu et al. in iSMART: ontology-based semantic query of CDA documents AMIA annual symposium, pp. 375–379, 2009) [6, 7]. HL7 specifies the structure and semantics of "clinical documents" for the purpose of sharing; therefore, data can be easily shared among the applications.

Keywords Cloud computing · Health cloud · HL7 RIM · Clinical document architecture · Health care as a service

A.K. Mandal (✉) · A. Sarkar
Department of Computer Applications, National Institute of Technology, Durgapur, India
e-mail: amitmandal.nitdgp@gmail.com

A. Sarkar
e-mail: sarkar.anirban@gmail.com

N. Chaki
Department of Computer Science and Engineering, University of Calcutta, Kolkata, India
e-mail: nabendu@ieee.org

© Springer India 2015   103
R. Chaki et al. (eds.), *Applied Computation and Security Systems*, Advances in Intelligent
Systems and Computing 304, DOI 10.1007/978-81-322-1985-9_8

1 Introduction

Technology today affects every single aspect of modern society. Hi-tech revolution affected every single sector, and its impact is more apparent in the field of medicine and health care than the other fields. Technological breakthroughs are revolutionizing the way health care is being delivered. Healthcare technologies have evolved over the years from departmental solutions to encompass larger solutions at the enterprise level, and from standalone systems that provide limited and localized solutions to more interconnected ones that provide comprehensive and integrated solutions. The complexity of healthcare system has also evolved from there being passive and reactive systems to now becoming more interactive and proactive with more focus on the quality of care [1]. Healthcare systems also benefited from the advancements in technology such as data storage systems, full redundancy for critical systems, as well as the recently emerging cloud computing environment to provide efficient and reliable solutions to support health services [2]. The widespread adoption cloud computing is changing modern health care and its delivery models at an ever-increasing rate. The inherent success of cloud-based service delivery models and the application has forced many researchers to rethink the often-debated question: can and how healthcare services be effectively offered through the cloud?

The requirements for computation power, storage, and continuous availability of healthcare data necessitate the use of the cloud computing services [3, 4]. It enhances the transfer, availability, and recovery of health records. It is also provides an easy and ubiquitous access to health data by improving medical services, opening new business models and opportunities. Although there is no standard definition of the healthcare cloud, it can be considered as a platform that, besides storing gigantic volumes of the health data, also serves as a structured management of the health data across multiple healthcare providers [5]. But the data in healthcare cloud are mostly semi-structured and may or may not conform to an open standard. These result in severe interoperability problems. In this regard, Health Level Seven (HL7) Reference Information Model (RIM) [6] and clinical document architecture (CDA) [7] can be considered as a standard data representation model. CDA is derived from RIM, and it is a document markup standard that specifies the structure and semantics of "clinical documents" for the purpose of exchange [7].

In healthcare domain, ontology and vocabulary standards enable semantic data integration, as they serve as a semantic reference for system programmers and users. It is also helpful in syntactic problems. HL7 CDA is intended to support the interchange of medical contents in context level. CDA is derived from RIM, which represents the information that has to be migrated into healthcare resource. HL7 RIM is developed based on UML. Act, Entity, and Role are defined as top-level classes in healthcare domain. In order to express one-to-many mapping relationship between instances of these key classes, three top classes, namely *"Participation," "Act_Relationship,"* and *"Role_link,"* are also defined [8]. Here, *"entity"* is physical things or group of physical things or organization, *"act"* defines a record of something that is being done or has been done, whereas a *"role"* is defined over an entity and defines its act. These classes are basic models that describe the CDA entry. Clinical documents, according to CDA, have six characteristics: persistence, stewardship, potential for authentication, context, wholeness, and human readability. CDA defines a header for classification and management and a document body that carries the clinical record. While the header metadata are prescriptive and designed for consistency across all instances, the body is highly generic.

Most of the cloud-based healthcare applications require secure, efficient, reliable, and scalable access to the medical records. Large numbers of medical records and images related to millions of people will be stored in healthcare clouds. The data may be replicated for high reliability and better access at different locations and across large geographic distances. Some of the data could also be made available locally [9]. These requirements enforce the need to have some storage services that provide fault tolerance, secure storage over public clouds, and rich query languages that allow efficient and scalable facilities to retrieve and process the application data [10]. In addition to these, the data in healthcare cloud must be consistent and constantly in a valid state regardless of any software, hardware, or network failures [11]. All cloud-based healthcare applications must deliver error-free services.

In this paper, a flexible cloud architecture for healthcare applications (FCAHCA) has been proposed for SaaS-based healthcare services, by incorporating all the features those are necessary for a cloud application such as scalability, flexibility, portability, adaptability, coexistence, and integrity. The proposal uses HL7 RIM and CDA standards [7]. The layered approach of the proposed architectural framework provides abstraction of different components in SaaS, suitable for the healthcare applications. It localizes the changes in one part of the solution and therefore minimizes the impact of change on the other parts. Moreover, it eases application maintenance and enhances overall application flexibility. Besides these, an extensive analysis of the proposed architectural framework is carried out using use cases and interaction diagrams to explain its functionality. The proposed architecture is a major extension of Flexible Cloud Architecture for Data-Centric Application (FCADCA) [12].

2 Related Works

In cloud computing paradigm, cloud service providers are not yet able to establish a standard methodology for designing, developing, and deploying cloud-based healthcare services. The published models and/or methodologies studied in the earlier literatures can be broadly classified into two categories based on the areas of interest: health cloud architectures and data storage in health cloud.

Recently, several researches have attempted toward the architectural model of health cloud, but only few of them are supportive toward flexible application development and heterogeneous medical data storage. Baru et al. [3] proposes CARE platform, the goal of which is to engender collaboration between the many domains of health information science and technology to leverage their respective strengths and unique capabilities. Lounis et al. [5] describes architecture for collecting and accessing large amount of data generated by medical sensor networks. It enables easy information sharing between healthcare professionals. Furthermore, it also describes an effective and flexible security mechanism that guarantees confidentiality, integrity, as well as fine-grained access control to outsourced medical data. Bahga et al. [4] propose an EHR system called cloud health information systems technology architecture (CHISTAR) that achieves semantic interoperability through the use of a generic design methodology. CHISTAR application components are designed using the cloud component model approach that comprises of loosely coupled components that communicate asynchronously. It also approaches for semantic interoperability, data integration, and security. Thuemmler et al. [13] proposes taxonomy and architecture for the implementation of the software-to-data paradigm in healthcare scenarios. The model is based on the "FI Core Platform." Ukis et al. [14] describes the key challenges involved in building a cloud-based solution for advanced medical image visualization and proposes a generic architecture that effectively addresses these challenges, whereas [15, 16] focuses on the design of a cloud framework for health monitoring system (CHMS). It collects patient's health data and publishes them to a cloud information repository. Further, it facilitates analysis of the data using services hosted in the cloud. However, most of these models are not properly supported with multitenancy, scalability, healthcare process management, semi-structured data management, data portability, etc. Moreover, multitenancy in cloud only supports the structured database operations and without having any metadata management services. In this regard, a scalable cloud architecture called Flexible Cloud Architecture for Data-Centric Applications (FCADCA) can be considered. It possesses several advantages over the others [17].

The health cloud application requires to deal with the large volume of data in different forms. Different cloud-based healthcare service provider uses their unique data representation and management techniques. Almutiry et al. [10] reviewed several cloud-based EHR systems and their functionalities, and based on this, it proposes a framework for data storage. Ermakova et al. [9] proposes a multiprovider cloud architecture that satisfies many of the requirements by providing increased availability, confidentiality, and integrity of the medical records stored in the cloud. This architecture features secret sharing as an important measure to distribute health records as fragments to different cloud services, which can provide higher redundancy and additional security and privacy protection in the case of key compromise, broken encryption algorithms, or their insecure implementation. Khan et al. [18] describe mechanisms to represent health regulations in machine-processable format. It also provides automatic reasoning about the compliances between the machine-processable regulations and the collected real-time data. Abbas et al. [11] surveyed the state-of-the-art privacy-preserving approaches employed in the e-Health clouds. Moreover, the privacy-preserving approaches have been classified into cryptographic and noncryptographic approaches and taxonomy of the approaches is also presented. Furthermore, the strengths and weaknesses of the presented approaches are reported and some open issues are highlighted.

Beside this, information exchange between several cloud service providers is equally important for health cloud applications. Mohammed et al. [19] outline a distributed Web interactive system for sharing health records on the cloud using distributed services and consumers, called Health Cloud eXchange. This system allows different health record and related healthcare services to be dynamically discovered and interactively used by client programs running within a federated private cloud. Others [20–22] propose ontology-based health record storage system in cloud environment. It may have a knowledge reasoning model which can include knowledge analysis module, ontology modeling module, decision engine, rule engine, and semantic template. The models provide a method which can analyze the diet and lifestyle information of people. And the information can be achieved from the domain ontology which is pretreated by domain experts. With the results of semantic reasoning processes, people can get some better healthcare services.

3 Scope of the Work

SaaS enhances collaboration among different healthcare organizations and to fulfill the common requirements, such as scale, agility, and cost-effectiveness. Besides several advantages of using SaaS for the delivery of healthcare services, it also offers some serious challenges which must be handled efficiently in order to establish SaaS as an effective service delivery model for healthcare services. These challenges can be categorized into business challenges, operational challenges, and technical challenges. Business challenge includes revenue model, payments, market, customer stickiness, and business models on SaaS. In operational domain service-level agreement (SLA), trust and quality of services are major concerns to be resolved.

Further, applications in public health and health services require access to a range of heterogeneous data, from environmental information in a region, to population-level data across regions, to more closely held personal health information, and from reference scientific data sets to observational data and sensor streams. Therefore, healthcare organizations are in the throes of a data explosion. Besides this, several incentives encourage hospitals everywhere to adopt electronic health records (EHRs). At the same time, there is an upsurge in the creation of Digital Imaging and Communications in Medicine (DICOM) data [8]. Therefore, numerous hospitals are implementing new picture archiving and communication systems (PACS) [8]. In addition to the aforementioned difficulties, in healthcare domain, 70–80 % of data are semi-structured in nature [23]. As a result, hospitals are finding it increasingly difficult and costly to manage all of these data. Even worse, most of the healthcare service providers are not meeting backup window requirements because there is simply too much data and not enough space and time for replication.

However, the most critical challenge toward providing healthcare services by SaaS model is majorly technical. *Firstly*, a cloud application needs to architect with reference to a standard architectural framework, but there is no common consensus in industry or academia on design and deployment of the cloud-based healthcare services. *Secondly*, most healthcare providers require high availability of the cloud-based healthcare services. Service and data availability is crucial for healthcare providers. It cannot effectively operate unless applications and patients' data are available [2]. Inside SaaS, the flexibility means the ease with which new services can be added or removed within a system without affecting the others. Thus, it is important that the environment of healthcare services lives in constant change and evolution. A healthcare service should be portable across the different service providers, and it should also be adaptable to any platforms or infrastructures [16]. *Thirdly*, a health cloud can contain several applications in a fully shared environment; therefore, they needed to coexist. Now, it is a common practice that a healthcare application is required to integrate with a third-party application on premises or hosted and there must be some mechanism to support it. *Finally*, on a daily basis, healthcare services deals with different kinds of digital information. This leads to semantic heterogeneity. Semantic heterogeneity occurs when there is a disagreement about the meaning, interpretation, or intended use of the same or related data. In this regard, usage of semi-structured data representation can be very effective as it contains tags or other markers to separate semantic elements and enforce hierarchies of records and fields within the data.

4 Flexible Cloud Architecture for Healthcare Applications

The proposed architectural framework called flexible cloud architecture for healthcare applications (FCAHCA) is devised using three core functionalities of SaaS, namely scalability, customizability, and multitenancy. Apart from these, the architecture is capable of providing the support for healthcare applications with interoperable data sets. Besides considering the standard functionalities of IaaS and PaaS in this architectural framework, an initiative has been taken to standardize the SaaS architecture for healthcare services. In this architecture, SaaS layer of healthcare cloud is divided into four basic layers, namely application layer, service management layer, healthcare process management layer, and data layer (Fig. 1). The data layer has been further divided into metadata management sublayer which uses HL7 RIM and CDA as standard for data representation, and CDA-compatible database management sublayer.

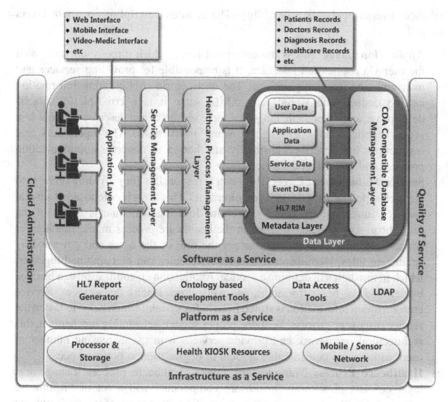

Fig. 1 Flexible cloud architecture for healthcare applications

In FCAHCA, each layer contains an *"interface"* which provides the means of interaction between different layers and it supports the exchange of information. Information exchange can be messages, which conforms a referenceable exchange pattern, structure, and semantics of the indicated information model. Interface also provides a barrier that enables the application logic to change without affecting the consumer of the interface. An interface may consist of a number of *"access points"* which provides access to the functionality exposed by the components of the layers. An access point of an interface can interact with more than one access point of the other interface across the different layers. The access points are selected dynamically for a particular service or service component based upon their availability and the type of the service or service components. Therefore, the easy addition or deletion of any service component for a particular service can be achieved.

The layered architecture is helpful toward the proper conceptualization and abstraction of different components of SaaS-based healthcare system. It also localizes the changes to one part of the solution and therefore minimizes the impact of changes on the other parts. Further, it eases application maintenance and enhances overall application flexibility. The detailed descriptions of these layers are as follows:

a) **Application Layer** This is the outermost layer, which directly interacts with the user via application interface. It is responsible for providing services and managing service configurations. The user-centric configuration information can be defined and stored with XML, since it is a cross-platform, self-descriptive, and extendable language. Here, designer can describe services as a set of applications operating on some instructions. The operations and instructions are described abstractly and bound to concrete communication protocol and offered as a services.

b) **Service Management Layer** This layer is responsible for providing services to the user as per their applications requirements. This layer is composed of several "service functional modules." A service functional module is a part of the "healthcare component" defined in the healthcare process management layer which performs a specific task as per the role of the users and governed by healthcare rules. Here, several functional modules can be combined together, which in turn helps in defining a service. Application customization can be done by defining the fields, formulas, and work flows among the services. Service customization information is stored in an index system, which helps to fasten the service discovery process. With the proper service-scheduling mechanism, this architectural framework can provide services with the increasing number of users without any considerable amount of degradation in performance.

c) **Healthcare Process Management Layer** Healthcare process management layer consists of several "healthcare component" and a set of "healthcare rules." A healthcare component is a set of healthcare process which may be considered atomic, and it is potentially fully automatable. Healthcare rules help in business modeling by enforcing some constraints in structure or in control behavior of the healthcare components. With this kind of framework, designers are able to achieve care-domain-oriented service grouping, which is independent of the functional modules used to define it. Therefore, the changes in this layer can be more readily accommodated since components and rules are separated. This provides higher flexibility toward the applications development. Based upon the defined healthcare rules, this layer can differentiate the level of access of the consumer over the application. It can also offer role-based service discovery and service policy by imposing some constraints. The components deployed in this layer can be shared among the different users or user groups. This layer is responsible for controlling the information exchange between service management layer and data layer.

d) **Data Layer** In this architectural framework, the data layer is divided into two sublayers, namely metadata management sublayer and semi-structured data management sublayer. Metadata management layer is based upon HL7 RIM ontology [6] and CDA [7] standards for data storage format and works as a bridge between healthcare process management sublayer and the CDA-compatible database management sublayer. Moreover, depending upon the types of data, metadata can be grouped as follows: "*user data*" which contains the user-related information; "*application data*" which holds application-related information; "*service data*" which is required for storing service properties and service indexing, whereas "*event data*" manages the healthcare events information. With metadata management layer, designers can separate the services and the related users logically.

The data layer of FCAHCA framework supports structured, semi-structured, and unstructured data representation, and storage of large-scale healthcare data. Semi-structured data are a form of structured data that do not conform to the data models associated with relational databases, but nonetheless contains tags or other markers to separate semantic elements and enforce hierarchies of records and fields within the data [24]. Therefore, the preponderance of semi-structured data generated by PACS, DICOM, and by enterprise content management (ECM) system can be managed more efficiently in this architectural framework.

Inherent in FCAHCA are characteristics that exacerbate existing quality-of-service (QoS) [25] concerns to the cloud environment. It is required for maintaining loose coupling, efficient composition of federated services, heterogeneous computing infrastructures, decentralized service-level agreements, and so on. These characteristics create complications for QoS which clearly require attention in any cloud environment. The QoS layer of FCAHCA enables noncompliance with service qualities as a whole and also in each layer of FCAHCA. Thus, the FCAHCA can capture, monitor, log, and signal noncompliance with nonfunctional requirements that relate to the service qualities. In some cases, the QoS layer can actually realize such nonfunctional requirements. In a sense, this layer observes the other layers and sends message or triggers events when it detects noncompliance, or preferably when it anticipates noncompliance. The QoS layer establishes issues related to nonfunctional requirements as a primary concern of cloud services and provides a focal point for dealing with them. It provides the means of ensuring that the architecture meets its requirements with respect to reliability, availability, manageability, scalability, and security. As such, the QoS layer is applicable to all other layers of FCAHCA architectural framework.

Moreover, the cloud administration layer covers all aspects of managing the cloud services. This layer monitors all policies from manual governance to cloud service policies. It can provision and create policies for cloud services and also monitoring violation of service-level agreements. Like QoS layer, this layer is also applicable to all other layers of FCAHCA.

5 Analysis of FCAHCA Using UML

The architecture of software system includes architectural concepts and appropriate syntax to represent them. The Unified Modeling Language (UML) [26] is the de facto standard for expressing the architectural design of software systems.

The UML modeling uses a notion view, which is a projection of the system on one of its relevant aspects. In this section, UML notation-based structural and behavioral views have been considered to analyze the functionality of proposed FCAHCA architectural framework.

UML provides several different notations which constitute different views of UML designs. Among these, use case diagram and sequence diagram are considered for analyzing the structural and behavioral aspects of any system. Those diagrams are useful toward analysis of functional aspects of the FCAHCA architectural framework. Use case diagrams display the relationship among actors and use cases. The use case depicts a set of scenarios describing interactions between users and components of FCAHCA. The analysis of FCAHCA using use case diagrams deals with set of interfaces and also with the behavioral characteristics at the boundary of the system. Again, sequence diagrams can be used to demonstrate the interaction of the components in FCAHCA. It primarily shows the sequence of events that occur. Sequence diagram therefore clearly defines the behavioral aspects of FCAHCA and are based on the structural and interface views.

The UML modeling helps in conceptualizing the different actions and semantics of the FCAHCA architectural framework in a more formal way. These actions and semantics of FCAHCA are represented using UML notations which are manifested by an abstract action language. Moreover, these modeling methodologies are useful to exhibit the detailed insight of the framework and effective toward the analysis and verification of functional characteristics of the proposed FCAHCA framework.

5.1 Use Case Representation of FCAHCA

Use case representation of FCAHCA illustrates possible valid engagements among its components; it also shows the capability of the system and demonstrates interaction mechanisms for the users of the system. In addition to this, a layer-specific analysis of FCAHCA is also carried out using use case diagrams. Figure 2 is a diagram which depicts the interaction among actors and different component and concepts of healthcare cloud. In this scenario, important facets of a FCAHCA architectural framework are described. Here, for simplicity, two different types of actors are considered "*service consumer*" and "*service provider.*"

As discussed in description of FCAHCA, service consumers directly interact with the application layer. The service consumer can specify "*service requirements*" and "*QOS parameters.*" The service consumer information along with the requirements is then passed to the service layer interface.

Service layer provides some mechanism which helps in "*search for services*" in "*service registry*" based on the user requirements. A consumer can "*select suitable services*" from the search results and subscribe to it by agreeing on "*service-level agreement*" (SLA). SLA defines a contract between a service provider and a service customer. Usually, it is specified in measurable terms about "what services

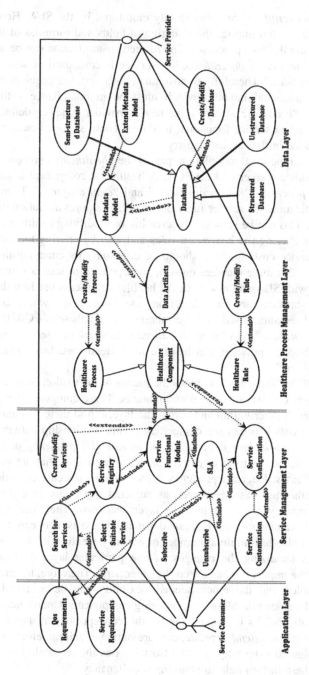

Fig. 2 Use case diagram for FCAHCA (only SaaS layer is considered)

the service provider will furnish." Service consumers may have different type of requirements; therefore, they may want to customize the services as per their needs. It may also able to unsubscribe a service. "*Service customization*" and service *"unsubscription"* are also strictly monitored by the SLA. Here, the term service customization implies the changing of fields and formulas of the services without altering the core processes. The service customization can be achieved by altering the *"service configurations."* A service is composed of several *"service functional modules."* Therefore, to customize a service, one requires customizing the service functional modules which must be in compliance with the SLA. However, to *"create or modify a service"* provider requires defining new or upgrading existing service functional modules. These service functional modules will be included in the service registry.

The service functional modules are part of the *"healthcare components,"* which is in the healthcare layer of FCAHCA. A healthcare component is composed of *"healthcare process," "healthcare rules,"* and *"data artifacts."* Therefore, creation or modification of a service functional module requires alteration in healthcare components. This can be achieved by creating or modifying healthcare process or by adding or altering the healthcare rules and also by altering the data artifacts. However, service customization should be done only by changing the rules and data artifacts. But these changes must be incorporated in such a way that it is in compliance with SLA and also with the healthcare process used by the services.

The data artifacts are extended from the *"metadata"* which confirms the structure and semantics used by the databases. The database of FCAHCA can be a *"structured database"* or a *"semi-structured database,"* or even an *"unstructured database."* Service provider can be able to "extend metadata" and *"create or modify database."*

In FCAHCA, *"interface"* provides the means of interaction between the layers and each layer consists of several components. These components require interaction with each other within and across the layers. And there is a many-to-many relationship exists between the components. This relationship between the components can be an *include* dependency or an *extend* dependency or aggregation. *Include* dependency is used to extract use case fragments that are duplicated in multiple use cases. *Extend* dependency is used when a use case conditionally adds steps to another use case. However, as depicted in the use case diagram, only *extend* dependency can exist in cross-layer scenarios. Therefore, in FCAHCA architectural framework, no *include* dependency exists between two adjacent layers, and it also has minimal number of *extend* dependencies across the layers. This enables localizing the changes in one particular layer of FCAHCA and minimizing the impact of changes on the other parts. Therefore, the architecture is loosely coupled and the much needed flexibility can be achieved using this architectural framework. Moreover, reducing components dependencies across the layers typically makes it easier to reuse the concepts of any layer in different context. Again, the *extend* dependencies are optional, and therefore, a component can be configured externally, which also increases the reusability of the components, and these in turn help in attaining multitenancy.

5.2 Interaction Among Different Components of FCAHCA

As discussed earlier, in FCAHCA, a layer interacts with the other layers via its interfaces. The end user interacts with the interface of the application layer. After a successful verification of the user and application information, this interface passes the application request to the interface of the service layer; otherwise, the application request is turned down (Fig. 3).

After a successful verification, the service-related information is checked for whether it matches with the capability of the service management layer or not. Here, service consumer can search for services by specifying the desired QOS values. And if the desired QOS values are not meet, then the service request is turned down. Service consumer can select suitable services from the search result. However, they can access that service only by agreeing on the SLA. The services may consist of several service functional modules. Consumers may also able to customize the services as per their requirements. This customization information is stored as service configuration. The service functional modules are accessed as per the service configuration (Fig. 4).

The service functional modules require accessing the healthcare components and healthcare rules and data artifacts in order to compose a healthcare service using FCAHCA; for this, it passes the healthcare process request through its interface to the healthcare process management layer. Here, configurations of healthcare components are checked to determine whether the requests are deliverable or not. Then, it invokes appropriate healthcare process instance and associates instances of the rule artifacts and data artifacts with the process instance and also propagates a request for the verification of the data artifacts to the metadata interface (Fig. 5).

The metadata management layer examines the request, and if the request meets the metadata configurations, then the metadata information becomes available to the healthcare process management layer. The metadata management layer then exchanges the information related to connection and storage with the database management layer interface. After verifying the connectivity, storage type and storage location database management layer establishes connection with the desired data storage system. The database can be an instance of a structured database or semi-structured database or an unstructured database (Fig. 6).

Here, successful access or retrieval of data from the database management layer confirms the metadata management layers request, as the database management layer is dependent on the metadata management layer. Again as discussed earlier, healthcare rule and data artifacts are using the metadata information; therefore, metadata management layers realize the healthcare process management layer. Dependency also exists between the service management layer and healthcare process management layer so the last layer's request approval confirms the first layer. In application layer, different service functional modules of service management layer are combined together and provided to the user as a full flagged

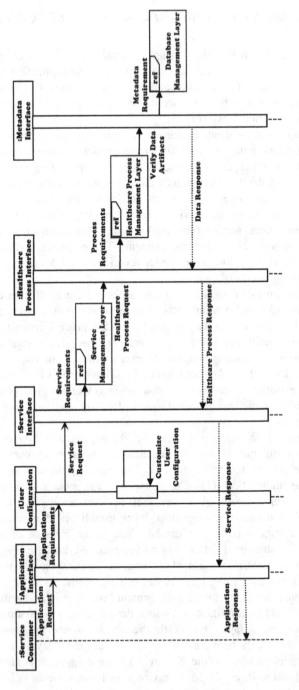

Fig. 3 Sequence diagram for FCAHCA (only SaaS layer is considered)

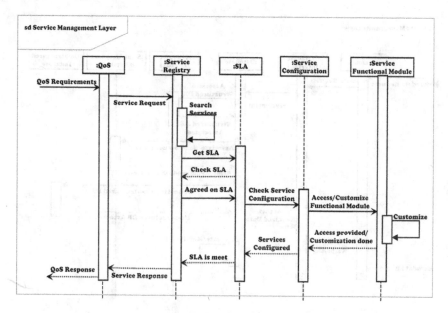

Fig. 4 Sequence fragments for service layer of FCAHCA

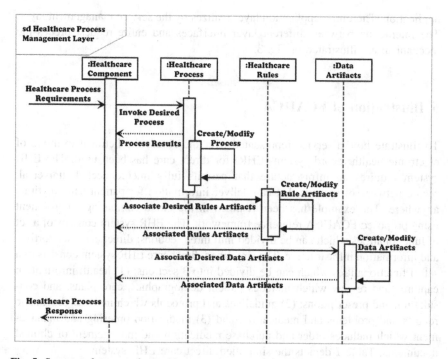

Fig. 5 Sequence fragments for healthcare process management layer of FCAHCA

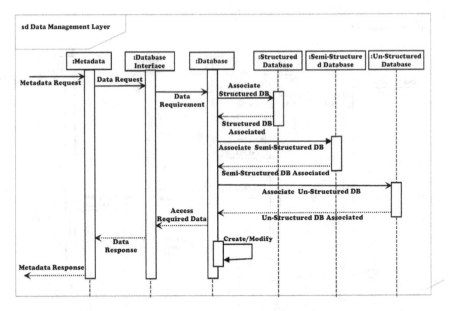

Fig. 6 Sequence fragments for database layer of FCAHCA

application. Therefore, application layer realizes by the service management layer. The interaction between different layer interfaces and entire request processing mechanism are illustrated in Fig. 3.

6 Illustration of FCAHCA

To illustrate how a service represented over the cloud, the fictional example of electronic health record system (EHR) for direct care has been used. The EHR system requires an infrastructure that permits fully interconnected, universal, secure network of systems that can deliver information for patient care anytime, anywhere. The example has been realized in the cloud computing environment using proposed FCAHCA reference framework. The EHR system consists of a set of functionalities which can be divided into three sections: direct care, supportive, and information infrastructure. The simplified direct care EHR system consists of a set of functionalities which can be divided into 3 sections: (1) health information captures and review which includes patient demographics, care plans, and consultation and prescriptions; (2) guidelines and protocols which includes diagnostic reports and problems and medication; and (3) medication ordering and management which includes order and purchase medication and management of clinical documents. Table 1 depicts the simplified direct care EHR system.

Table 1 Components of direct care EHR system

Direct care EHR domain	ID	Direct care EHR components
Health information capture and review	HC 1	Patient demographic
	HC 2	Guidelines
	HC 3	Consultation and prescriptions
Guidelines and protocols	HC 4	Diagnostic reports
	HC 5	Problem and medication
Medication ordering and management	HC 6	Order and purchase medication
	HC 7	Manage clinical documents

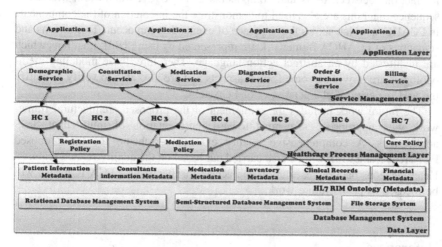

Fig. 7 Direct care EHR system using FCAHCA

As shown in the Fig. 7, the healthcare process management layer of direct care EHR system is composed of several healthcare components as listed in Table 1. This layer also contains a set of healthcare rules such as registration policy, medication policy, and care policy. Registration policy regulates the registration process, whereas care policy introduces some constraints depending upon the patients' medical insurance. These healthcare rules are combined with healthcare components. In this, illustration service management layer consists of services like demographic service, consultation service, medication service, etc. The service in this layer uses some specific healthcare components. Here, the same type of services may be grouped together. In application layer, the services are combined together to form an application and provided as a service. Moreover, in direct care EHR system, metadata is defined by HL7 RIM and CDA. The metadata layer can be grouped as patient information metadata, consultation metadata, medication metadata, inventory metadata, clinical records metadata, and financial metadata based on their structure and semantics. Application data are stored in the EHR database which is capable of storing structured, semi-structured, and unstructured types of data.

7 Conclusion

In this paper, a new SaaS architectural framework for cloud-based healthcare applications (FCAHCA) has been proposed. This architectural framework is capable of handling semi-structured data sets and supports separation of concerns among its components. It also possesses the several key characteristics of SaaS those are necessary to develop and deploy cloud-based healthcare applications.

In this paper, a detailed analysis of the FCAHCA architectural framework also has been illustrated using UML, specifically by using use case diagram and sequence diagram. The use case analysis depicts a set of scenarios describing interactions between users and components of FCAHCA framework. Moreover, sequence diagrams are also used to demonstrate the interaction of the components in FCAHCA, which shows the sequence of events that can occur. This actions and semantics of FCAHCA framework are represented by using UML notations which are manifested by an abstract action language. Therefore, these modeling methodologies are providing some formal aspects and are also able to verify or clarify the vast majority of characteristics of FCAHCA.

The future work includes the refinement of the proposed architectural framework with appropriate semantics toward representation of multitenant SaaS applications. A comprehensive quality evaluation scheme for the proposed FCAHCA framework will be also carried out based on the quality-related features. Heterogeneous data representation and access mechanism for FCAHCA also will be focused as prime future work.

References

1. Yun, J.H., Kim, L.K.: Processing HL7-CDA entry for semantic interoperability, International Conference on Convergence Information Technology, pp. 1939–1944 (2007)
2. AbuKhousa, E., Mohamed, N., Al-Jaroodi, J.: e-Health cloud: Opportunities and challenges. Future Internet 4, 621–645 (2012). doi:10.3390/fi4030621
3. Baru, C., Botts, N., Horan, T.: A seeded cloud approach to health cyberinfrastructure: Preliminary architecture design and case applications, IEEE 45th Hawaii International Conference on System Sciences, pp. 2727–2734 (2012)
4. Bahga, A., Madisetti, V.K.: A cloud-based approach for interoperable electronic health records (EHRs). IEEE J. Biomed. Health Inf. 17(5), 894–906 (2013)
5. Lounis, A., Hadjidj, A., Bouabdallah, A., Challal, Y.: Secure and scalable cloud-based architecture for e-health wireless sensor networks. In: IEEE 21st International Conference on Computer Communications and Networks (ICCCN), pp. 1–7 (2012)
6. Hennessy, M., Oentojo, C., Ray, S.: A framework and ontology for mobile sensor platforms in home health management, URL: www.sei.cmu.edu/community/mobs2013/.../ MOBS2013-Hennessy.pdf (2013)
7. Liu, S., Ni, Y., Mei, J., Li, H., Xie, G., Hu, G., Liu, H., Hou, X., Pan, Y.: iSMART: Ontology-based semantic query of CDA documents AMIA Annual Symposium, pp. 375–379 (2009)
8. Cotterill, T.: So much data, so little time, URL: http://www.healthmgttech.com/articles/ 201011/so-much-data-so-little-time.php (2010)

9. Ermakova, T., Fabian, B.: Secret sharing for health data in multi provider clouds. In: IEEE International Conference on Business Informatics, pp. 93–100 (2013)
10. Almutiry, O., Wills, G., Alwabel, A., Crowder, R., Walters, R.: Toward A Framework for Data Quality in Cloud-Based Health Information System, pp. 153–157 (2013)
11. Abbas, A., Khan, S.U.: A Review on the state-of-the-art privacy preserving approaches in the e-health clouds. In: IEEE Journal of Biomedical and Health Informatics. DOI: 10.1109/JBHI. 2014.2300846, (In Press)
12. Mandal, A.K., Changder, S., Sarkar, A., Debnath, N.C.: A novel and flexible cloud architecture for data-centric applications, IEEE International Conference on Industrial Technology (ICIT), pp. 1834–1839 (2013)
13. Thuemmler, C., Magedanz, T., Jell, T., Mueller, J., Covaci, S., de Panfilis, S., Schneider, A., Gavras, A.: Applying the software-to-data paradigm in next generation E-Health hybrid clouds. In: IEEE 10th International Conference on Information Technology: New Generations, pp. 459–463 (2013)
14. Ukis, V., Balachandran, B., Tirunellai Rajamani, S., Friese, T.: Architecture of cloud-based advanced medical image visualization solution In: IEEE International Conference on Cloud Computing in Emerging Markets (CCEM), pp. 1–5 (2013)
15. Reddy, B.E., Kumar, T.V.S., Ramu, G.: An efficient cloud framework for health care monitoring system. In: International Symposium on Cloud and Services Computing (ISCOS), vol. 17, 18, pp. 113–117 (2012). doi:10.1109/ISCOS.2012. http://ieeexplore.ieee.org/stamp/stamp.jsp?tp=&arnumber=6481246&isnumber=6481216
16. Rashid, Z., Farooq, U., Jang, J.K., Park, S.H.: Cloud computing aware ubiquitous health care system. In: IEEE 3rd International Conference on E-Health and Bioengineering, pp. 1–4 (2011)
17. Mandal, A.K., Changder, S., Sarkar, A., Debnath, N.C.: Architecting software as a service for data centric cloud applications. In: Accepted and will Appear in International Journal of Grid and High Performance Computing (IJGHPC), vol. 6, Issue 1. IGI-Global, USA (2014)
18. Kha, K.M., Bai, Y.: Automatic verification of health regulatory compliance in cloud computing. In: IEEE 15th International Conference on e-Health Networking, Applications and Services, pp. 713–717 (2013)
19. Mohammed, S., Servos, D., Fiaidhi, J.: HCX: A distributed OSGi based web interaction system for sharing health records in the cloud. In: IEEE/WIC/ACM International Conference on Web Intelligence and Intelligent Agent Technology, pp. 102–107 (2010)
20. Deng, H., Zhang, X., Liu, J.: Knowledge reasoning in health cloud. In: IEEE International Conference on Cloud and Service Computing, pp. 48–54 (2011)
21. Bhuvaneswari, A., Karpagam, G.R.: Ontology-Based emergency management system in a social cloud. Int. J. Cloud Comput.: Serv. Archit. (IJCCSA). 1(3), 15–29 (2011)
22. Chuan-Jun, Su, Chiang, Chang-Yu.: IAServ: An intelligent home care web services platform in a cloud for aging-in-place. Int. J. Environ. Res. Public Health 10(11), 6106–6130 (2013)
23. Kilic, Ozgur, Dogac, Asuman: Achieving clinical statement interoperability using R-MIM and archetype-based semantic transformations. IEEE Trans. Inf Technol. Biomed. 13(4), 467–477 (2009)
24. Sarkar, A.: Design of semi-structured database system: conceptual model to logical representation, In: Singh, H., Kaur, K.(eds.) Book Titled: Designing, Engineering, and Analyzing Reliable and Efficient Software, IGI Global Publications, USA
25. Arsanjani, A., Zhang, L.J., Ellis, M., Allam, A., Channabasavaiah, K.: S3: A service-oriented reference architecture. IT Professional. 9(3), pp. 10–17 (2007)
26. OMG Unified Modeling Language TM (OMG UML), Superstructure, Version 2.4.1, Document Number: formal/2011-08-06, URL: http://www.omg.org/spec/UML/2.4.1/Superstructure, Aug 2011

Part III
Wireless Sensor Networking

An Overlay Cognitive Radio Model Exploiting the Polarization Diversity and Relay Cooperation

Sandip Karar and Abhirup Das Barman

Abstract Interference mitigation is one of the significant challenges in overlay cognitive radio (CR). This paper proposes a scheme to avoid the interference in an overlay CR scenario by exploiting the polarization diversity and relay cooperation. In this scheme, a dual-polarized antenna is incorporated at the secondary transmitter, one of which acts as a relay to the primary user's signal and cancels out the interference via space–time coding. Analytical results about the error performance and the outage probability for the scheme have been calculated. The performance degradation in the absence of perfect CSI has also been shown.

1 Introduction

With rapid growth of wireless applications, the problem of spectrum utilization has become much more critical. Cognitive radio (CR) technology offers a promising solution, in which the licensed spectrum can be reused in three ways, namely *underlay, overlay,* and *interweave* method [1]. In *underlay* and *overlay* methods, both the primary and the secondary users can simultaneously use the same spectrum, whereas in *interweave* method, the secondary user continually searches for the spectrum holes and uses those spectrum holes for communication. Different interference mitigation techniques in *overlay* CR network have been proposed such as MIMO precoding [2], beamforming [3, 4], etc. In most of the overlay techniques, it has been assumed that the secondary transmitter has non-causal information about the primary user's message or the full channel state information (CSI) which is quite impractical.

S. Karar (✉) · A. Das Barman
Department of Radiophysics and Electronics, University of Calcutta, Kolkata, India
e-mail: sk7632@gmail.com

A. Das Barman
e-mail: abhirup_rp@yahoo.com

© Springer India 2015
R. Chaki et al. (eds.), *Applied Computation and Security Systems*, Advances in Intelligent Systems and Computing 304, DOI 10.1007/978-81-322-1985-9_9

In this paper, we exploit the polarization diversity to mitigate the interference in an overlay spectrum sharing scheme. The use of the polarization of the radio signal is not new in wireless communication, but only a few papers have addressed the issue [5–9]. But exploiting the polarization of the signal in the context of inter-ference mitigation in overlay CR has not been done previously to the best of our knowledge. Our scheme requires a dual-polarized antenna at the secondary trans-mitter side, one of which acts as a decode-and-forward (DF) relay to the primary and cancels out the interference via space–time coding and the other secondary antenna is used for secondary transmission. Only the receivers need to know the CSI of all the links. The channel gains can be estimated by pilot symbols before the actual communication starts. This scheme certainly reduces the throughput of the primary user. But if the direct line-of-sight communication link of the primary user is comparatively worse than the link via secondary relay, then this scheme would be immensely useful. A scheme similar to this has been previously proposed in [10] where the secondary transmitter uses two separate antennas to form a MISO system. One of the main disadvantages of that scheme is that the antennas must be separated large enough resulting in larger size and higher cost of the secondary user. This scheme is replaced here by a dual-polarized antenna as a space and cost-effective alternative. Also larger number of channels needs to be estimated beforehand [10] in comparison with the dual-polarized scheme used here.

The rest of the paper is organized as follows: Sect. 2 gives an analysis of a 2×2 CR model where the primary and the secondary users share the spectrum using polarization multiplexing. In Sect. 3, we propose our relay based commu-nication scheme using dual-polarized antenna at the secondary, and analytical formulation of average symbol error rate and outage probability for the scheme have been carried out. Sect. 4 gives the analysis for average symbol error rate and outage probability of the same scheme under imperfect CSI. Simulation results are shown in Sect. 5, and conclusions are drawn in Sect. 6.

2 A 2×2 Cognitive Radio Model Using Orthogonal Polarization

A schematic model for a 2×2 overlay CR has been shown in Fig. 1 where the primary users (both Tx and Rx) are using vertically polarized antennas and the secondary users are using horizontally polarized antennas. The symbols h_{pp}, h_{ps}, h_{sp} and h_{ss} denote the flat Rayleigh fading channel gains each with parameter σ_h. In ideal scenario, the cross-polar transmission, i.e., transmission from vertically polarized Tx antenna to horizontally polarized Rx antenna or vice versa, should be zero. But in reality, there is always some cross-polar leakage mainly due to the rotation of the plane of polarization when the signal propagates through the atmosphere.

It has been shown in [9] that the leakage from vertically to horizontally polarized transmission and horizontally to vertically polarized transmission has the

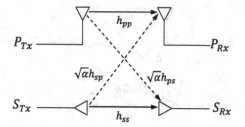

Fig. 1 A 2 × 2 cognitive radio model using polarization multiplexing

same power on average. Let α be the fraction of the cross-polar leakage power where $0 < \alpha \leq 1$. To incorporate the cross-polar leakage, the channel gain between orthogonal polarized antennas should be multiplied by $\sqrt{\alpha}$. Let the primary transmitter transmit the symbol x_p with power P_p and the secondary transmitter transmit the symbol x_s with power P_s simultaneously. We assume $E(|x_p|^2) = 1$ and $E(|x_s|^2) = 1$. The received signals at the primary and secondary receivers are given by

$$y_p = \sqrt{P_p}h_{pp}x_p + \sqrt{\alpha P_s}h_{sp}x_s + n_p \tag{1}$$

$$y_s = \sqrt{P_s}h_{ss}x_s + \sqrt{\alpha P_p}h_{ps}x_p + n_s \tag{2}$$

Here, n_p and n_s are the additive white Gaussian noise with zero mean and variance N_0 at the primary and secondary receivers, respectively. In ideal scenario $\alpha = 0$, the term x_s vanishes in the expression of y_p and the term x_p vanishes in the expression of y_s. So there would be no interference at each receiver. But in actual case, $\alpha \neq 0$ due to cross-polar leakage, and hence, this causes some interference both to the primary and secondary receivers. In the next section, we will show how our proposed scheme can mitigate the interference completely in overlay CR.

3 Proposed Scheme for Interference Mitigation for a 2 × 2 Cognitive Radio

It is obvious from the previous section that using the orthogonal polarization between two users cannot remove the interference completely due to cross-polar leakage. Here, we propose a scheme exploiting the polarization diversity and relay cooperation for a 2 × 2 overlay CR model that can take care of the cross-polar leakage and mitigate the interference completely. The model is similar as before; only difference is that instead of one horizontally polarized antenna, the secondary transmitter is equipped with a dual-polarized antenna: one horizontally and one vertically.

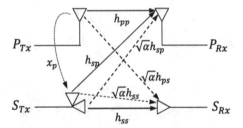

Fig. 2 A 2×2 cognitive radio model using relay cooperation and space–time coding

It is assumed that both the receivers know the CSI of all the links. The secondary transmitter serves as a DF relay to the primary signal. The system model is shown in the Fig. 2. The transmission takes place in two phases. In the first phase, the primary transmitter P_{Tx} transmits the signal x_p with power P_p and the secondary transmitter S_{Tx} transmits the signal x_s with power P_s through its horizontally polarized antenna. Simultaneously, the vertically polarized antenna of the secondary transmitter also receives the primary transmission and decodes it. In the second phase, the primary transmitter remains silent, but the secondary transmitter transmits the signal x_p^* with power P_p through its vertically polarized antenna and the signal $-x_s^*$ with power P_s through its horizontally polarized antenna. In the first transmission phase, the signal received by the primary and secondary receivers, respectively, are given by

$$y_{p1} = \sqrt{P_p}h_{pp}x_p + \sqrt{\alpha P_s}h_{sp}x_s + n_{p1} \tag{3}$$

$$y_{s1} = \sqrt{P_s}h_{ss}x_s + \sqrt{\alpha P_p}h_{ps}x_p + n_{s1} \tag{4}$$

and in the second transmission phase, the signal received by the primary and secondary receivers, respectively, are given by

$$y_{p2} = \sqrt{P_p}h_{sp}x_p^* - \sqrt{\alpha P_s}h_{sp}x_s^* + n_{p2} \tag{5}$$

$$y_{s2} = -\sqrt{P_s}h_{ss}x_s^* + \sqrt{\alpha P_p}h_{ss}x_p^* + n_{s2} \tag{6}$$

where n_{p1}, n_{p1}, n_{p2}, and n_{p2} are additive white Gaussian noise with zero mean and variance N_0. Similar to Alamouti's scheme [11] of space–time coding, the receivers can now properly combine the received signals in the two-phase transmission to mitigate the interfering terms if perfect CSI is available at the receivers. From (3), (4), (5), and (6), we can write:

$$h_{sp}^* y_{p1} + h_{sp}y_{p2}^* = \sqrt{P_p}\left(h_{pp}h_{sp}^* + |h_{sp}|^2\right)x_p + h_{sp}^* n_{p1} + h_{sp}n_{p2}^* \tag{7}$$

$$h_{ss}^* y_{s1} - h_{ps}y_{s2}^* = \sqrt{P_s}\left(|h_{ss}|^2 + h_{ss}^* h_{ps}\right)x_s + h_{ss}^* n_{s1} - h_{ps}n_{s2}^* \tag{8}$$

From these expressions, it is evident that the interference at both the primary and the secondary receivers is completely removed. However, the interference rejection at the secondary receiver depends on how well the secondary transmitter can estimate the transmission power of the primary. Here, we assume perfect estimation, but a slight wrong estimation can lead to substantial interference at the secondary receiver. The signal-to-noise ratio at the primary receiver and secondary receiver are then respectively given by $\gamma_p = \frac{P_p \left| h_{pp} h_{sp}^* + \left| h_{sp} \right|^2 \right|^2}{2 \left| h_{sp} \right|^2 N_0}$ and $\gamma_s = \frac{P_s \left| \left| h_{ss} \right|^2 + h_{ss}^* h_{ps} \right|^2}{\left(\left| h_{ss} \right|^2 + \left| h_{ps} \right|^2 \right) N_0}$. Assuming Rayleigh distribution of the channels, it can be derived that the average received primary SNR $\overline{\gamma_p} = \frac{3 P_p \sigma_h^2}{N_0}$, where σ_h is the parameter of the Rayleigh distributed fading channels. Similarly, it can be shown $\overline{\gamma_s} = \frac{3 P_s \sigma_h^2}{N_0}$. When M-PSK modulation is used, the symbol error probability for the primary user is given by [12]

$$P_{se}(\text{primary}) \approx \alpha_M Q \left(\sqrt{\beta_M \gamma_p} \right) \tag{9}$$

where $\alpha_M = 2$ and $\beta_M = 2 \sin^2 \left(\frac{\pi}{M} \right)$.

The average symbol error probability can be calculated as

$$\overline{P_{se}}(\text{primary}) \approx \int_0^\infty \alpha_M Q \left(\sqrt{\beta_M \gamma_p} \right) p(\gamma_p) \mathrm{d}\gamma_p \tag{10}$$

From (10), it can be derived [12]:

$$\overline{P_{se}}(\text{primary}) = \frac{\alpha_M}{2} \left[1 - \sqrt{\frac{0.5 \beta_M \overline{\gamma_p}}{1 + 0.5 \beta_M \overline{\gamma_p}}} \right] = \frac{\alpha_M}{2} \left[1 - \sqrt{\frac{1.5 \beta_M P_p \sigma_h^2}{N_0 + 1.5 \beta_M P_p \sigma_h^2}} \right] \tag{11}$$

The outage probability P_{out} is defined as the probability that the received SNR falls below a given threshold corresponding to the maximum allowable symbol error probability. The outage probability of the primary user relative to the threshold γ_{p0} is given by

$$P_{\text{out}}(\text{primary}) = P(\gamma_p \leq \gamma_{p0}) = \int_0^{\gamma_{p0}} \frac{1}{\overline{\gamma_p}} e^{-\gamma_p / \overline{\gamma_p}} \mathrm{d}\gamma_p = 1 - e^{-\gamma_{p0} / \overline{\gamma_p}}$$

$$= 1 - e^{-\frac{\gamma_{p0} N_0}{3 P_p \sigma_h^2}} \tag{12}$$

In the similar way, the average symbol error probability and the outage probability relative to the threshold γ_{s0} for the secondary user can be calculated.

$$\overline{P_{se}}(\text{secondary}) = \frac{\alpha_M}{2}\left[1 - \sqrt{\frac{1.5\beta_M P_s \sigma_h^2}{N_0 + 1.5\beta_M P_s \sigma_h^2}}\right] \tag{13}$$

$$P_{\text{out}}(\text{secondary}) = 1 - e^{-\gamma_{s0}\overline{\gamma_s}} = 1 - e^{-\frac{\gamma_{s0}N_0}{3P_s\sigma_h^2}} \tag{14}$$

From (11), (12), (13), and (14), it is obvious that the error performance and outage probability of this scheme are completely independent of α, i.e., the cross-polar leakage power ratio.

4 Modeling for Imperfect CSI and Analysis for Average Symbol Error Probability and Outage Probability

Let g be the estimate of the channel h so that we can write $h = g + d$, where d is the zero mean Gaussian estimation error. The channel h and the estimate of the channel g are also zero-mean Gaussian random variables. According to principal of orthogonality, the optimal LMSE yields to an estimation error orthogonal to the channel realization h [13]. Then, we can write $\sigma_h^2 = \sigma_g^2 + \sigma_d^2$, where σ_h^2, σ_g^2 and σ_d^2 are the variance of the real part (or imaginary part) of h, g and d, respectively. So for the above scheme, we can rewrite the Eqs. (7) and (8) as

$$g_{sp}^* y_{p1} + g_{sp} y_{p2}^* = \sqrt{P_p}\left(h_{pp}g_{sp}^* + h_{sp}^* g_{sp}\right)x_p + \sqrt{\alpha P_s}\left(h_{sp}g_{sp}^* - h_{sp}^* g_{sp}\right)x_s \\ + g_{sp}^* n_{p1} + g_{sp}n_{p2}^* \tag{15}$$

$$g_{ss}^* y_{s1} - g_{ps} y_{s2}^* = \sqrt{P_s}\left(h_{ss}g_{ss}^* + h_{ss}^* g_{ps}\right)x_s + \sqrt{\alpha P_p}\left(h_{ps}g_{ss}^* - h_{ss}^* g_{ps}\right)x_p \\ + g_{ss}^* n_{s1} - g_{ps}n_{s2}^* \tag{16}$$

So the interference at both the primary and the secondary receivers is not completely removed due to the imperfect CSI which leads to performance degradation. From (15), we can write

$$g_{sp}^* y_{p1} + g_{sp} y_{p2}^* = \sqrt{P_p}\left((g_{pp} + d_{pp})g_{sp}^* + (g_{sp}^* + d_{sp}^*)g_{sp}\right)x_p \\ + \sqrt{\alpha P_s}\left((g_{sp} + d_{sp})g_{sp}^* - (g_{sp}^* + d_{sp}^*)g_{sp}\right)x_s \\ + g_{sp}^* n_{p1} + g_{sp}n_{p2}^* \\ = \sqrt{P_p}\left(g_{pp}g_{sp}^* + |g_{sp}|^2\right)x_p \\ + \sqrt{P_p}\left(d_{pp}g_{sp}^* + d_{sp}^* g_{sp}\right)x_p \\ + \sqrt{\alpha P_s}(d_{sp}g_{sp}^* - d_{sp}^* g_{sp})x_s + g_{sp}^* n_{p1} + g_{sp}n_{p2}^* \tag{17}$$

From this expression, we can extract the message part, interference part due to imperfect CSI, and the noise part, respectively, as follows:

$$M = \sqrt{P_p}\left(g_{pp}g_{sp}^* + |g_{sp}|^2\right)x_p \tag{18a}$$

$$I = \sqrt{P_p}\left(d_{pp}g_{sp}^* + d_{sp}^*g_{sp}\right)x_p + \sqrt{\alpha P_s}(d_{sp}g_{sp}^* - d_{sp}^*g_{sp})x_s \tag{18b}$$

$$N = g_{sp}^* n_{p1} + g_{sp} n_{p2}^* \tag{18c}$$

The signal-to-interference plus noise ratio at the primary receiver is then given by

$$\gamma_p = \frac{P_p\left|g_{pp}g_{sp}^* + |g_{sp}|^2\right|^2}{P_p|d_{pp}g_{sp}^* + d_{sp}^*g_{sp}|^2 + \alpha P_s|d_{sp}g_{sp}^* - d_{sp}^*g_{sp}|^2 + 2|g_{sp}|^2 N_0} \tag{19}$$

In the similar way, the signal-to-interference plus noise ratio at the secondary receiver can be calculated as follows:

$$\gamma_s = \frac{P_s\left||g_{ss}|^2 + g_{ss}^*g_{ps}\right|^2}{P_s|d_{ss}g_{ss}^* + d_{ss}^*g_{ps}|^2 + \alpha P_p|d_{ps}g_{ss}^* - d_{ss}^*g_{ps}|^2 + (|g_{ss}|^2 + |g_{ps}|^2)N_0} \tag{20}$$

From the expression of γ_p and γ_s, we can calculate the average signal-to-interference plus noise ratios as $\overline{\gamma}_p = \frac{3P_p(\sigma_h^2 - \sigma_d^2)}{(2P_p + \alpha P_s)\sigma_d^2 + N_0}$ and $\overline{\gamma}_s = \frac{3P_s(\sigma_h^2 - \sigma_d^2)}{2(P_s + \alpha P_p)\sigma_d^2 + N_0}$. The average symbol error probabilities can be calculated as follows:

$$\overline{P}_{se}(\text{primary}) = \frac{\alpha_M}{2}\left[1 - \sqrt{\frac{0.5\beta_M\overline{\gamma}_p}{1 + 0.5\beta_M\overline{\gamma}_p}}\right]$$

$$= \frac{\alpha_M}{2}\left[1 - \sqrt{\frac{1.5\beta_M P_p(\sigma_h^2 - \sigma_d^2)}{(2P_p + \alpha P_s)\sigma_d^2 + N_0 + 1.5\beta_M P_p(\sigma_h^2 - \sigma_d^2)}}\right] \tag{21}$$

$$\overline{P}_{se}(\text{secondary}) = \frac{\alpha_M}{2}\left[1 - \sqrt{\frac{0.5\beta_M\overline{\gamma}_s}{1 + 0.5\beta_M\overline{\gamma}_s}}\right]$$

$$= \frac{\alpha_M}{2}\left[1 - \sqrt{\frac{1.5\beta_M P_s(\sigma_h^2 - \sigma_d^2)}{2(P_s + \alpha P_p)\sigma_d^2 + N_0 + 1.5\beta_M P_s(\sigma_h^2 - \sigma_d^2)}}\right] \tag{22}$$

Similarly, we can now also calculate the outage probabilities of the primary and the secondary users relative to the corresponding thresholds γ_{p0} and γ_{s0}, respectively, as follows:

$$P_{\text{out}}(\text{primary}) = 1 - e^{-\gamma_{p0}/\overline{\gamma_p}} = 1 - e^{-\frac{\gamma_{p0}((2P_p+\alpha P_s)\sigma_d^2+N_0)}{3P_p(\sigma_h^2-\sigma_d^2)}} \tag{23}$$

$$P_{\text{out}}(\text{secondary}) = 1 - e^{-\gamma_{s0}/\overline{\gamma_s}} = 1 - e^{-\frac{2\gamma_{s0}((P_s+\alpha P_p)\sigma_d^2+N_0)}{3P_s(\sigma_h^2-\sigma_d^2)}} \tag{24}$$

5 Simulations and Discussion

The analytical expressions derived above have been plotted in MATLAB. We set the values of the simulation parameters such as the parameter of the Rayleigh faded channels $\sigma_h^2 = 0.7$ and the threshold for outage $\gamma_{p0} = \gamma_{s0} = 2\,\text{dB}$. For simulation purpose, we keep the power of the secondary transmitter equal to the primary transmission power and α value is taken to be 0.1. In Fig. 3, the average symbol error probability or symbol error rate (SER) performance of the primary user is plotted with respect to the signal-to-noise ratio (P_p/N_0) for perfect CSI and different fixed values of error-to-actual channel variance ratio (ER) which is defined as $\text{ER} = \frac{\sigma_d^2}{\sigma_h^2}$. Here, we use $M = 4$, i.e., QPSK modulation for simulation. Figure 4 shows the average symbol error rate plot for the secondary user with the same specification. Since we have assumed $P_p = P_s$, the plots of Figs. 4 and 5 show nearly similar nature as evident from (21) and (22).

Figures 5 and 6 show the plots of outage probability at the primary and secondary receivers, respectively, relative to the corresponding threshold $\gamma_{p0} = \gamma_{s0} = 2\,\text{dB}$ with respect to the signal-to-noise ratio for perfect CSI and for different ER values which was drawn using the analytical formula (23) and (24). From the plots, it is obvious that the perfect CSI provides the best performance for both the symbol error rate and the outage probability, and as the ER value is increased, the symbol error rate and the outage probability performance tend to worsen and asymptotically converge toward a lower limit and it becomes impossible to achieve anything below that limit even by increasing the signal-to-noise ratio arbitrarily. This lower-limit value rises as the ER value is increased.

And finally, Figs. 7 and 8 show the spectrum efficiency plot of the primary and secondary users, respectively, with respect to the signal-to-noise ratio for perfect CSI and for different ER values and comparison with the TDMA scheme. In TDMA scheme, equal and orthogonal time slots are allotted to the primary and secondary users. Our proposed scheme exhibits slight superior performance in terms of spectral efficiency over TDMA scheme. In case of imperfect CSI, the

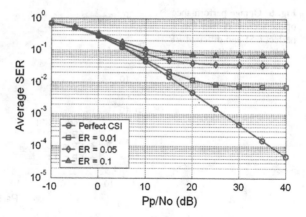

Fig. 3 Symbol error performance of the primary user for perfect CSI and different ER values

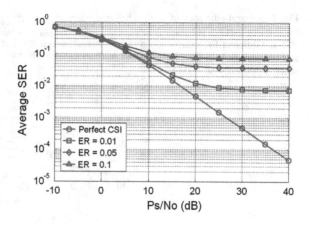

Fig. 4 Symbol error performance of the secondary user for perfect CSI and different ER values

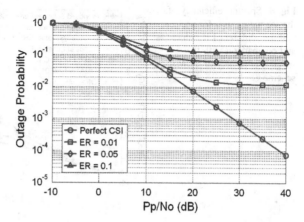

Fig. 5 Outage performance of the primary user for perfect CSI and different ER values ($\gamma_{p0} = 2\,\text{dB}$)

Fig. 6 Outage performance of the secondary user for perfect CSI and different ER values ($\gamma_{s0} = 2\,\text{dB}$)

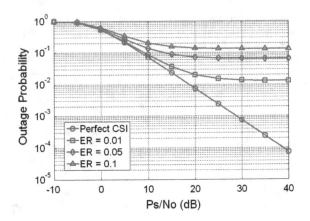

Fig. 7 Spectral efficiency of the primary user for perfect CSI and different ER values and comparison with the spectral efficiency of TDMA scheme

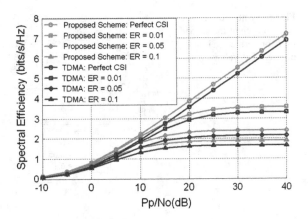

Fig. 8 Spectral efficiency of the secondary user for perfect CSI and different ER values and comparison with the spectral efficiency of TDMA scheme

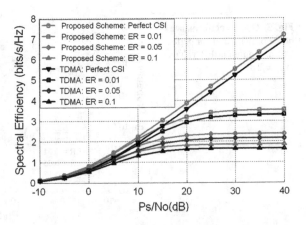

spectral efficiency does not go on increasing with the SNR but converge toward an upper limit both for our proposed scheme and TDMA scheme. This upper limit value decreases as ER value is increased.

6 Conclusions

In this work, we have proposed an interference mitigation scheme of a 2×2 overlay CR model using the polarization diversity and relay cooperation. The error performance and the outage performance for this scheme have also been shown. One main advantage of this scheme is that under perfect CSI, the error performance and outage probability are completely independent of the cross-polar leakage power ratio. It has also been shown that the absence of perfect CSI can cause severe performance degradation of the system. The simulation results show very little dependence on the cross-polar leakage power ratio on the system performance even in the absence of perfect CSI.

Acknowledgments The work is undertaken as part of Media Lab Asia project entitled "Mobile Broadband Service Support over Cognitive Radio Networks."

References

1. Sudhir Srinivasa, S., Jafar, S.A.: The throughput potential of cognitive radio: a theoretical perspective. In: Fortieth Asilomar Conference on Signals, Systems and Computers (2006)
2. Chen, Z., Wang, C.-X., Hong, X., Thompson, J., Vorobyov, S.A., Zhao, F., Xiao, H., Ge, X.: Interference mitigation for cognitive radio MIMO systems based on practical precoding, Arxiv preprint arXiv:1104.4155, vol. abs/1104.4155 (2011)
3. Zhang, L., Liang, Y.C., Xin, Y.: Joint beamforming and power control for multiple access channels in cognitive radio networks. IEEE J. Sel. Areas Commun. **26**(1), 38–51 (2008)
4. Yiu, S., Vu, M., Tarokh, V.: Interference reduction by beamforming in cognitive networks. In: Proceedings of IEEE Global Communication Conference (GLOBE-COM), pp. 1–6 (2008)
5. Vaughan, R.: Polarization diversity in mobile communications. IEEE Trans. Veh. Technol. **39**(3), 177–186 (1990)
6. Nabar, R.U., Bölcskei, H., Erceg, V., Gesbert, D., Paulraj, A.J.: Performance of multiantenna signaling techniques in the presence of polarization diversity. IEEE Trans. Signal Process. **50**(10), 2553–2562 (2002)
7. Erceg, V., Soma, P., Baum, D.S., Catreux, S.: Multiple-input multiple-output fixed wireless radio channel measurements and modeling using dual-polarized antennas at 2.5 GHz. IEEE Trans. Wireless Commun. **3**(6), 2288–2298 (2004)
8. Oestges, C., Clerckx, B., Guillaud, M., Debbah, M.: Dual-Polarized wireless communications: from propagation models to system performance evaluation. IEEE Trans. Wireless Commun. **7**(10), 4019–4031 (2008)
9. Coldrey, M.: Modeling and capacity of polarized MIMO channels. In: IEEE Vehicular Technology Conference, Singapore (2008)

10. Bohara, V.A., Ting, S.H., Han, Y., Pandharipande, A.: Interference-free overlay cognitive radio network based on cooperative space time coding. In: 5th International Conference on Cognitive Radio Oriented Wireless Networks and Communications, Cannes (2010)
11. Alamouti, S.M.: A simple transmitter diversity scheme for wireless communications. IEEE J. Sel. Areas Commun. **16**, 1451–1458 (1998)
12. Goldsmith, A.: Wireless communications. Cambridge University Press, Cambridge (2005)
13. Seyfi, M., Muhaidat, S., Liang, J.: Amplify-and-forward selection cooperation over Rayleigh fading channels with imperfect CSI. IEEE Trans. Wireless Commun. **11**(1), 199–209 (2012)

Effect of Source Selection, Deployment Pattern, and Data Forwarding Technique on the Lifetime of Data Aggregating Multi-sink Wireless Sensor Network

Kaushik Ghosh, Pradip K. Das and Sarmistha Neogy

Abstract Data aggregation has been used as a prominent technique for lifetime enhancement of wireless sensor networks (WSN) for quite some time. Data aggregation reduces total number of transmissions in a WSN. Since transmitting energy is the most prominent component of energy consumption in a WSN, data aggregation reduces energy expenditure of the network and thereby enhances network lifetime. The nature of aggregation, however, may vary from one application to another. Along with this, the way source nodes are selected for transmission has an effect on the energy depletion and lifetime of the nodes. In this paper, we have studied the effect of certain non-electrical factors such as source selection, deployment pattern, packet size, and data forwarding technique on the performance of aggregation of a multi-sink WSN with varying degrees of aggregation.

Keywords Data aggregation · Network lifetime · Source selection mode · Deployment pattern · Data forwarding technique · Packet size

1 Introduction

Lifetime enhancement of wireless sensor networks (WSN) through reduction in energy expenditure has been a prominent topic of research for over a decade. Of the different techniques proposed by the researchers for the same, data aggregation

K. Ghosh (✉)
Mody University of Science and Technology, Laxmangarh, Sikar, Rajasthan, India
e-mail: kaushikghosh.fet@modyuniversity.ac.in

P.K. Das
RCCIIT, Kolkata, West Bengal, India
e-mail: pkdas@ieee.org

S. Neogy
Jadavpur University, Kolkata, West Bengal, India
e-mail: sarmisthaneogy@gmail.com

© Springer India 2015   137
R. Chaki et al. (eds.), *Applied Computation and Security Systems*, Advances in Intelligent Systems and Computing 304, DOI 10.1007/978-81-322-1985-9_10

is one that has been talked about for quite sometime [1, 2]. Of the different factors determining energy consumption of a WSN, energy consumed for transmission is the most prominent one and affects lifetime of a WSN more than any other components [3]. Since through data aggregation it is possible to reduce total number of transmissions, it eventually reduces the energy consumption of the network [4]. Aggregation is done at certain points in the network where data from multiple sources are collected and multiple input packets are reduced to one output packet and transmitted to the sink(s). Since energy expenditure of a WSN is dependent upon the total number of bits transmitted and the distance travelled by a packet [3, 5], data aggregation is capable of reducing energy expenditure of a network by n folds, where n incoming packets are processed and converted into a single outgoing packet [4]. This way energy saving is done by (i) reducing the number of transmissions as well as (ii) reducing the total number of bits to be transmitted n folds as compared to the number of bits received. Radio models proposed in [3, 5] however, confirm that of the two factors, (i) inter-nodal distance and (ii) number of bits to be transmitted, the former consumes much more energy as compared to the latter. Processing n incoming packets and converting them into a single outgoing packet demand some amount of computational energy to be spent. But since the amount of computational energy ($E_{computational}$) is negligble compared to energy required for transmission or reception [6], it is worth taxing the processor of a sensor node instead of its transceiver. The aggregating nodes accept n input packets and perform aggregation functions such as average, max, min, duplicate suppression to have a single output packet. For example, if the application under consideration is for measuring the average temperature or humidity of a given area, then it is redundant to transmit readings from every node deployed. Rather, finding the average of a subset of nodes at some aggregation point and forwarding a single packet to the sink (s) is the sensible option. In certain applications, however, every single datum is of importance. For example, if sensors are attached to the body of patients in a hospital to monitor parameters such as heart beat and blood pressure, then data from every individual source are important and not a mere average or maximum value. In scenarios like this, aggregator cannot convert n input packets to one output packet. Still in this case considerable amount of energy can be saved as compared to cases with no aggregation by reducing the number of transmissions alone—n input packets may be clubbed together into a single bigger output packet of size n times that of the individual input packets. Unlike the previous scheme, total number of bits transmitted in this case, however, remains the same as the total number of bits received. Here, the number of bits in the header of the single larger packet would be considerably less than the number of bits in n different headers. In this paper, we have thus considered two scenarios: (a) data aggregation that converts n input packets to one of size $1/n$ of the input ones and (b) data aggregation that converts n input packets to one of size n times the input ones. We then compared the lifetime of a WSN for different aggregation factors and different methods of source node selection. We also illustrate that lifetime of a WSN may be further enhanced over the presence of aggregation by deciding upon certain non-electrical factors such as deployment pattern, source selection mode, and data forwarding technique. In this paper, we have compared the lifetime of a WSN under different source selection modes, deployment patterns, and data forwarding techniques. Aggregation factor determines the degree of aggregation. It is the number of input packets to be aggregated into one output packet. The rest of the paper is organized as follows: in Sect. 2, we have discussed some related works, Sect. 3 contains our proposed scheme, Sect. 4 contains results, and the conclusion appears in Sect. 5.

2 Related Works

Krishnamachari et al. [4] proposed a data centric approach of routing in place of traditional address centric approach to address the energy constrained nature of WSNs. The paper examined the impact of source-destination placement and network density on the energy costs and delay associated with data aggregation. In fact, impact of network density upon data aggregation was discussed in [7] as well. Here, greedy aggregation was compared with opportunistic approach while finding mean dissipated energy for varying number of neighboring nodes. The findings confirmed that in high-density networks, the greedy approach achieves significant energy saving as compared to opportunistic approach. Massad et al. [8] tells about multiple initiators for each data-gathering iteration and claims that their proposed approach will consume energy in a uniform way when compared with serial data aggregation schemes. The work in [9] presents an exact as well as approximate algorithm to find the minimum number of aggregation points in order to maximize the network lifetime. The approximate algorithm produces results not very different from the exact one. The work in [10] proposed a novel aggregation scheme that adaptively performed application-independent data aggregation. The aggregation decisions are kept in a module between network and data link layer and do not require any modification on the existing protocols of either layer. The protocol reduces end to end delay by 80 % and transmission energy consumption by 50 %. He et al. [11] present two privacy-preserving data aggregation schemes for additive aggregation functions. The objective of the authors was to bridge the gap between collaborative data collection by WSN and data privacy.

Data aggregation reduces the number of transmissions that in turn reduces energy consumption and thereby increases network lifetime. Another method of increasing lifetime is to reduce the total transmitting distance. For that, over past few years authors have used Fermat point-based routing protocols. The work in [12] discusses how routing distance can be improved through a Fermat point-based scheme. A farther improvement over results of [12] became possible with the inclusion of the concept of inside relay nodes (INR) [13], when it came to (i) total squared Euclidian distance and (ii) energy consumption.

Effect of network lifetime on forwarding technique selection has been discussed by Ghosh and Das [14]. Here, the authors have compared a Fermat point-based data forwarding technique with greedy forwarding and residual energy-based forwarding. Network lifetime was recorded maximum for the Fermat point-based scheme as compared to the other two schemes.

It thus becomes evident that a protocol involving both data aggregation and Fermat point-based forwarding is likely to consume less energy as compared to schemes which include either of the two above-mentioned approaches. The advantage of selecting Fermat point as the aggregation point is shown in [15]. The advantage of selecting Fermat point as aggregation point is—unlike other aggregation schemes, it does not require any tree to be constructed for aggregation. The authors compared the Fermat point-based scheme with greedy incremental tree (GIT) forwarding and the results obtained showed enhanced lifetime for the Fermat point-based scheme as compared to GIF. Lifetime here was recorded by varying network density. An extension of the same work was carried out in [16]. Here, a decentralized hierarchical aggregation scheme using Fermat point was proposed. Like Son et al. [15], here also GIT forwarding was compared with the proposed Fermat point-based scheme. The parameters considered here were number of hops for transmission and number of working nodes after certain rounds of transmission. In both the cases, the Fermat point-based scheme outperformed GIT scheme.

A detailed survey of different lifetime enhancement techniques for WSNs has been proposed in [17]. Here, along with the algorithms that involve battery management techniques for lifetime enhancement in WSNs, Fermat point-based algorithm for lifetime enhancement was also discussed. Of the different protocols discussed here, almost all of them (other than [14]) consider movement of data from a single source to a single destination. The work in [18] discusses a novel Fermat point-based data dissemination scheme which constructs a disseminating tree from one root (Fermat point) to multiple leaves (sinks) and thus addresses the issue of multiple sinks in a WSN. For performance evaluation, the authors compared their scheme with two-tier data dissemination and delay-constrained minimum energy dissemination. Energy consumption using the scheme discussed in [18] was found to be less as compared to the other two schemes for both control and data packets.

3 Proposed Scheme

In the previous section, we have discussed how a Fermat point-based aggregation scheme is useful in reducing energy consumption, thereby enhancing network lifetime. This motivated us to propose a Fermat point-based aggregating and forwarding scheme for multi-sink WSNs.

The WSN under consideration is assumed to be comprised of multiple sinks. Sensor nodes are deployed randomly over a two-dimensional polygonal plane with m vertices. $m - 1$ sinks are placed at those many vertices of the said polygon. The network follows a Fermat point-based forwarding technique [19, 20]. Fermat point is defined as the point within the bounds of a polygon or triangle (with no internal angle greater than 120°) such that the sum of the straight lines from all the vertices to that point is minimum. The theoretical Fermat point for the polygonal region formed by taking the source and sinks is first found out. Then, the physical node closest to the theoretical Fermat point is found out. We will call this node the *Fermat node* (FN) for the said source. Thus, every source would have a FN of its own. A node acting as source at one moment may act as FN at some other time for some other node. A node may also act as FN for multiple sources. A source routes a packet to its FN, and then, it is the responsibility of FN to forward the packet to different sinks. After the packet reaches different sinks, it is the turn of some other nodes to act as the source. The packet is thus forwarded from the source to the sinks in two phases: firstly, packet is routed to the FN — its intermediate destination, and secondly the FN forwards the packet to the sinks. In both phases, the packet is assumed to reach the destination/intermediate destination in either single or multiple hops. Figure 1 depicts the above-mentioned scenario. In this paper, we have discussed the effect of factors such as (i) node deployment pattern, (ii) source selection mode, (iii) data forwarding technique and (iv) packet size on the lifetime of a WSN. For applications such as military surveillance or arid region weather monitoring, the sensor nodes are deployed randomly—possibly by air dropping—as direct human intervention may either be impossible or is not desirable. But there are certain other application areas where nodes may be deployed exactly at the desired location, e.g., precision agriculture, health care. We have thus compared network lifetime for two different scenarios—in one, nodes are deployed randomly, and in the other, they are deployed in a grid fashion. The source selection mechanism may be either random or round robin in nature. In round robin mechanism, nodes are selected as source serially according to their node ids, one after another. Thus, for a network of n nodes, once a node acts as source, it does so again after all of the remaining $n - 1$ nodes have played their part as source nodes. Random source selection on the other hand does not guarantee anything of that sort. Here, a node acting as source may again be selected as the source node before other nodes got selected as source, at least once. In Sect. 4, we will see how different modes of source selection (random and round robin) affects network lifetime. Once a source node is selected, it is now up to the source

Fig. 1 Source node, Fermat node, and sinks

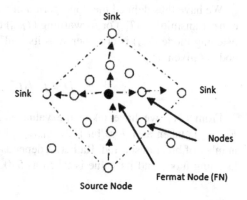

node to decide a suitable forwarding technique to forward data to one of its neighbors. Here, we have compared variants of greedy forwarding technique, compass routing, and residual energy-based forwarding with a composite parameter-based forwarding technique proposed in [14]. In greedy forwarding technique, a node forwards data to the neighbor who is closest to the destination. Compass routing selects the neighbor that makes minimum angle with the straight line joining destination node with the node currently holding data packet. In residual energy-based forwarding, the residual energy of a node is taken under consideration while selecting forwarding node. It may be mentioned here that the greedy forwarding, compass routing, and residual energy-based forwarding are all Fermat point-based. First, data is transferred from source to FN and then from FN to different sinks. Data aggregation is done at FN only.

According to the forwarding technique proposed in [14], a node forwards a packet to the neighbor with the highest value for *forwarding potential*. The forwarding potential (κ) of any node is calculated as

$$\kappa = \text{res\_energy}/\text{dist} \tag{1}$$

where
res_energy =   Residual energy of a node in milli Joules
dist =   Distance of a node from a particular sink in meters

The major sources of energy consumption considered in this paper are sensing (E_{Sensing}), computation ($E_{\text{computation}}$), forwarding ($E_{\text{forwarding}}$), receiving ($E_{\text{receiving}}$), and listening ($E_{\text{listening}}$). Energy consumed by the source node (E_{TX}) comprises the components—E_{Sensing}, $E_{\text{computation}}$, and $E_{\text{forwarding}}$. Relay nodes on the other hand need to receive a packet before they can further forward it. Thus, $E_{\text{receiving}}$ and $E_{\text{forwarding}}$ are the components of energy consumption for relay nodes. We demarcate energy consumption of the relay nodes as $E_{\text{forwarding}}$. Finally, we assume that when a node neither acts as a sender nor a relay, then it is listening to the transmissions of its neighbors. The energy consumed for the said purpose is given as $E_{\text{listening}}$. A node is considered to be in *on* state, while it acts as either sender or relay. On the other hand, while on listening phase, it is said to be in its *off* state.

We have thus defined the "on" period of a node (t_{on}) as the time it is engaged in either transmitting (T_{tx}) or forwarding (T_{fwd}) data. The time for which a node is in listening mode (T_{lst}) is considered as its "off" period (t_{off}). The duty cycle D of a node is given as

$$D = t_{\text{on}}/(t_{\text{on}} + t_{\text{off}}) \tag{2}$$

From [21], we have taken the value of $E_{\text{computation}}$ as 117 nJ/bit. Similarly, E_{sensing} is taken as 1.7 μJ/bit [22]. $E_{\text{listening}}$ on the other hand is not a function of number of bits transmitted. Rather, it depends upon the number of seconds spent in listening mode and its value is taken as 570 μJ/s [23].

So, the radio model used in this paper is of the following form:

$$E_{TX} = m \times 117 \times 10^{-9} + m \times 1.7 \times 10^{-6} + D \times m \times \varepsilon \times d^n \tag{3}$$

$$E_{forwarding} = D \times (m \cdot E + m \times \varepsilon \times d^n) \tag{4}$$

$$E_{listening} = (1 - D) \times 570 \times 10^{-6} \tag{5}$$

where

$m =$ Packet size in number of bits
$D =$ Duty cycle
$E =$ 50 nJ/bit
$\varepsilon =$ 8.854 pJ/bit/m^2
$d =$ Inter-nodal distance

The values of E and ε are taken from [3]. ε stands for permittivity, and E is the minimum start-up energy required for any communication.

In the proposed scheme, we assign the responsibility of data aggregation on the nodes acting as FN (Fig. 1). Data aggregation imposes some extra load on the processor and thus consumes some computational energy ($E_{computational}$). Thus, the nodes involved with the task of data aggregation are expected to deplete their energy faster as compared to the nodes who are not assigned the said task. From Fig. 1, we understand that the nodes located around the periphery of the polygonal region have less or no chance to be selected as FN. Rather, it is the nodes in the middle who are selected more often as FNs for one node or the other.

A node selected as the FN is thus assigned with two responsibilities: (i) forwarding packets to the sinks and (ii) aggregating data. On receiving a packet, anode acting as FN would increment a counter and compare the value of the same with the *aggregation factor* (AGFACT) decided earlier. Since $E_{computational}$ is negligible as compared to transmission or reception energy, more would be the energy saving for a higher value of aggregation factor. This is, as discussed earlier, due to lesser number of transmissions required by the FN when compared to a scheme not involving data aggregation. Figure 2 depicts the scenario how number of transmissions are reduced when we take FN as the aggregation point (AG-FACT = 2). The scenario depicted in Fig. 2b makes the Fermat node wait for packets from both the nodes N1 and N2 before forwarding the aggregated data to the three sinks. Thus, we see the number of transmissions has reduced to three as compared to six in Fig. 2a where no aggregation is used. Furthermore, it is evident from the figures that more the aggregation factor more would be the savings in energy due to reduced number of transmissions, and thus, the lifetime of the network will also increase. For lifetime calculation, we have used the definition proposed by the authors in [24], i.e., the network is considered dead, when the first node in it goes out of energy. However, as we keep on increasing the AGFACT, the waiting time in FN would increase and that would increase the delay. There has

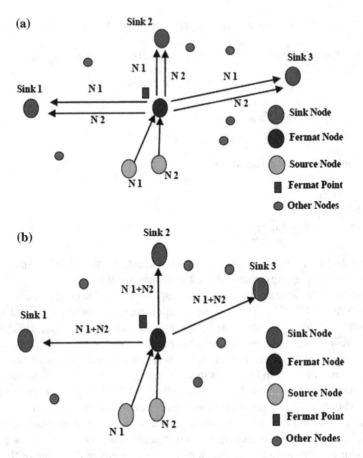

Fig. 2 a Scheme not involving data aggregation. **b** Scheme involving data aggregation (AGFACT = 2)

to be a trade-off between the increased lifetime and the delay induced. In the next section, we have shown how increasing the AGFACT would increase network lifetime along with increase in delay. Moreover, we have compared the network lifetime for random and round robin type of source selection to see their effects on a particular network.

4 Results

This section comprises four subsections—one each for discussing the effect of source selection, deployment pattern, forwarding strategy, and packet size. Sensor nodes are considered to be deployed within a rectangle with sinks at three of the vertices of the same rectangle. Two subsequent transmissions from node i and node j are assumed to be separated by a transmission interval of 1 hour. The said scenario is simulated using C (gcc compiler).

4.1 Effect of Source Selection

For studying the effect of different source selection modes, the set of parameters used is shown in Table 1. As discussed earlier, we have considered two different aggregation schemes. If p is the size in bits of each input packet, then the two aggregation schemes are (a) data aggregation that converts n input packets to one packet of size p and (b) data aggregation that converts n input packets to one packet of size $n \times p$. The number of transmissions involved at the aggregation points for both the schemes remains the same. But as the number of bits transmitted per transmission is less for scheme (a), it is obvious that this scheme would consume less energy as compared to (b) at the aggregation points. Therefore, the lifetime of an application applying type (a) aggregation is recorded more as compared to type (b) for both round robin and random source selection modes. This is recorded in Fig. 3a, b, respectively. Although scheme (a) yields more lifetime compared to (b), the increment, as we can see from Fig. 3a, b, is not considerable. This, we infer, is due to the fact that the component of energy consumption involving number of transmissions is much more dominant over the component of energy consumption involving number of bits to be transmitted. This again is in accordance with the radio models proposed in [4, 3]. From Fig. 3c, we find that round robin source selection mode outperforms the random one for both aggregation scheme (a) and (b). In both the cases, we have kept the aggregation factor as equal to 4.

4.2 Effect of Deployment Pattern

To monitor the effect of two different deployment patterns, viz. (a) random and (b) grid deployment, the network parameters were chosen as shown in Table 2. Figure 4a–c shows how deployment pattern (b) outperforms (a) in terms of network

Table 1 Parameters selected for different source selection modes

Parameters	Values
Nodes	100
Area	150 m × 150 m
Number of sinks	3
Deployment pattern	Random
Transmission range (TXR)	80 m
Data rate	38.4 kbps
Packet size	36 bytes
Initial energy of the nodes	1 J

Fig. 3 **a** Lifetime
comparison with increase in
aggregation factor using
round robin source selection.
b Lifetime variation with
increase in aggregation factor
using random source
selection. **c** Lifetime
comparison for round robin
and random source selection
modes

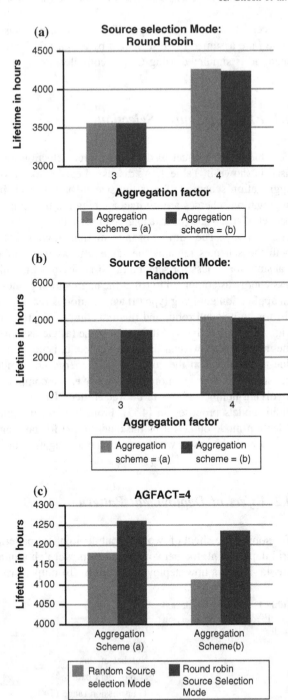

Table 2 Parameters selected for different deployment patterns

Parameters	Values
Source selection mode	Random
Transmission range	80–100 m
Aggregation scheme	(b)

Fig. 4 **a** Lifetime comparison for grid and random deployment (TXR = 80 m). **b** Lifetime comparison for grid and random deployment (TXR = 90 m). **c** Lifetime comparison for grid and random deployment (TXR = 100 m)

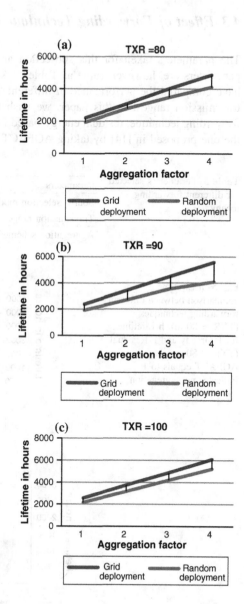

lifetime when all other parameters remain same. For grid deployment, we have segregated the 150 m × 150 m rectangular area into smaller grids with cells of size 15 m × 15 m each and have placed a sensor node at the center of each cell.

4.3 Effect of Forwarding Technique

The parameters taken for this subsection are same as Table 3. The remaining parameters are, however, same as Table 1. As discussed earlier, in [14], authors have compared the performance of different forwarding techniques with varying transmission ranges. In this paper, we wish to compare performance of greedy forwarding technique, residual energy-based forwarding and compass routing with the one proposed in [14] by taking AGFACT as the independent variable.

Table 3 Parameters selected for different forwarding techniques

Parameters	Values
Source selection mode	Random
Transmission range	80 m
Aggregation scheme	(b)

Fig. 5 **a** Lifetime comparison between different forwarding techniques (TXR = 80 m). **b** Lifetime for F-Greedy and F-Residual (TXR = 80 m). *Note* AGFACT equals to 1 indicates no aggregation

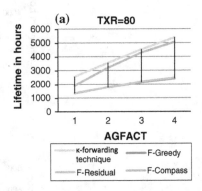

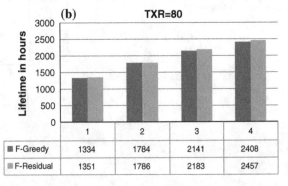

Table 4 Parameters selected for varying packet sizes

Parameters	Values
Number of nodes	100
Source selection	Random
Deployment pattern	Random
Forwarding technique	κ-Forwarding technique
Aggregation scheme	(b)
Data rate	38.4 kbps
Initial energy of nodes	1 J
Number of sinks	3
Transmission range	80 m
Area	150 m $\times$ 150 m

Let us name the forwarding technique of [14] as κ-*forwarding technique*, as per the demarcation of the forwarding potential. Figure 5a, b show that for different aggregation factors, lifetime of a network is more when κ-forwarding technique is used over three other forwarding schemes: Fermat point-based Greedy Forwarding (F-Greedy), Fermat point-based Residual Energy-based Forwarding (F-Residual) and Fermat point-based Compass Routing (F-Compass).

4.4 Effect of Packet Size

For this section, we have selected the parameters according to Table 4.

Here, we have recorded the lifetime variation of a network for different packet sizes with different degrees of aggregation. The variable *PCKTSZE* in Fig. 6 accounts for packet sizes in the multiple of 36 bytes. For a given AGFACT, network lifetime reduces with increase in packet size.

Fig. 6 Lifetime variation for increasing packet size

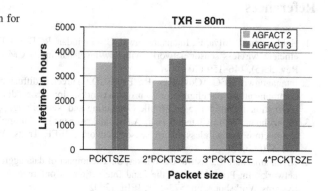

5 Conclusion

With all electrical specifications remaining same, lifetime of a network with multiple sinks may increase considerably when proper care is taken while choosing certain non-electrical parameters like deployment pattern, source selection mode, and forwarding technique. Due to random deployment of nodes, internodal distances between different nodes remain non-uniform. This in turn causes non-uniform energy depletion in the nodes, and thus, lifetime of the network recorded is lower as compared to the one where node deployment has been done in a uniform manner, e.g., grid deployment.

Similarly, when source selection mode is random, nodes may be selected as source more frequently as compared to round robin source selection mode. This again leads to premature energy depletion in nodes in case we use random source selection mode. Forwarding data using F-Greedy forwarding method again is not a very sensible choice in WSNs. Topology of the network being static, for a particular source–sink pair, same nodes are used over and over again for forwarding data. F-Residual forwarding too leads to premature depletion of energy for nodes with higher energy as compared to nodes in its vicinity, as it becomes the natural choice for its neighbors as the forwarding node for a particular sink. However, it performs marginally better compared to greedy forwarding since after a node's energy level falls below that of its neighbors; it is no longer selected as the forwarding node and thus is saved from complete energy depletion.

F-Compass forwarding too is outperformed by κ-forwarding technique, but it has given better results as compared to the other two forwarding mechanisms.

More the packet size, less is the lifetime of a network for a given AGFACT. This in fact is in accordance with our radio model.

Finally, we conclude that on comparison with κ-forwarding technique, the other three techniques register lower lifetime for a network for a given transmission range and aggregation factor.

References

1. Heidemann, J., Silva, F., Intanagonwiwat, C., Govindan, R., Estrin, D., Ganesan, D.: Building efficient wireless sensor networks with low-level naming. ACM SIGOPS Operating Syst. Rev. **35**(5), 146–159 (2001)
2. Intanagonwiwat, C., Govindan, R., Estrin ,D.: Directed diffusion: a scalable and robust communication paradigm for sensor networks. In: Proceedings of the 6th Annual International Conference on Mobile Computing and Networking, pp. 56–67. ACM (2000)
3. Heinzelman, W.B., Chandrakasan, A.P., Balakrishnan, H.: An application-specific protocol architecture for wireless micro sensor networks. IEEE Trans. Wireless Commun. **1**(4), 660–670 (2002)
4. Krishnamachari, L., Estrin, D., Wicker, S.: The impact of data aggregation in wireless sensor networks. In: Proceedings of the 22nd International Conference on Distributed Computing Systems Workshops, pp. 575–578. IEEE (2002)

5. Hwang, I.-S., Pang, W.: Energy efficient clustering technique for multicast routing protocol in wireless adhoc networks. IJCSNS 7(8), 74–81 (2007)
6. Min, R., Bhardwaj, M., Cho, S.-H., Shih, E., Sinha, A., Wang, A., Chandrakasan, A.: Low-power wireless sensor networks. In: International Conference on VLSI Design, pp. 205–210 (2001)
7. Intanagonwiwat, C., Estrin, D., Govindan, R., Heidemann, J.: Impact of network density on data aggregation in wireless sensor networks. In: Proceedings of the 22nd International Conference on Distributed Computing Systems (ICDCS'02), pp. 457–458 (2002)
8. Massad, Y.E., Goyeneche, M., Astrain, J.J., Villadangos, J.: Data aggregation in wireless sensor networks. In: 3rd International Conference on Information and Communication Technologies: From Theory to Applications, ICTTA 2008, pp. 1–6. IEEE (2008)
9. Al-Karaki, J.N., Ul-Mustafa, R., Kamal, A.E.: Data aggregation in wireless sensor networks—exact and approximate algorithms. Workshop on High Performance Switching and Routing, 2004, HPSR, pp. 241–245. IEEE (2004)
10. He, T., Blum, B.M., Stankovic, J.A., Abdelzaher, T.: AIDA: adaptive application-independent data aggregation in wireless sensor networks. ACM Trans. Embed. Comput. Syst. (TECS) 3(2), 426–457 (2004)
11. He, W., Liu, X., Nguyen, H., Nahrstedt, K., Abdelzaher, T.T.: PDA: privacy-preserving data aggregation in wireless sensor networks. In: 26th IEEE International Conference on Computer Communications, INFOCOM 2007, pp. 2045–2053. IEEE (2007)
12. Ssu, K.F., Yang, C.H., Chou, C.H., Yang, A.K.: Improving routing distance for geographic multicast with Fermat points in mobile ad hoc networks. Comput. Netw. 53(15), 2663–2673 (2009)
13. Long Chen, Y., Jun Ding, W., Chi Chang, Y., Chung Wang, N.: Applications for improving geographic routing paths in wireless sensor networks. J. Adv. Comput. Netw. 1(4), 334–338 (2013)
14. Ghosh, K., Das, P.K.: Effect of forwarding strategy on the lifetime of multi-hop multi-sink sensor network. In: Third International Conference on Trends in Information, Telecommunication and Computing, vol. 150. LNEE (2013)
15. Son, J., Pak, J., Han, K.: Determination of aggregation point using Fermat's point in wireless sensor networks. APWeb Workshops 2006, vol. 3842. LNCS, pp. 257–261 (2006)
16. Son, J., Pak, J., Kim, H., Han, K.: A decentralized hierarchical aggregation scheme using Fermat points in wireless sensor networks. Evo Workshops 2007, vol. 4448. LNCS, pp. 153–160 (2007)
17. Saraswat, J., Rathi, N., Bhattacharya, P.P.: Techniques to enhance lifetime of wireless sensor networks: a survey. Global J. Comput. Sci. Technol. Netw. Web Secur. 12(14), version 1.1, 21–31 (2012)
18. Chuang, P.-J., Li, B.-Y.: Fermat point based data dissemination in sensor networks. J. Chin. Inst. Eng. 32(7), 959–966 (2009)
19. Lee, S.H., Ko, Y.B.: Geometry-driven scheme for geocast routing in mobile adhoc networks. In: IEEE Conference on Vehicular Technology, 2006. IEEE, pp. 638-642 (2006)
20. Ghosh, K., Roy, S., Das, P.K.: An alternative approach to find the Fermat point of a polygonal geographic region for energy efficient geocast routing protocols: global minima scheme. In: First International Conference on Networks and Communications, NetCoM 2009. IEEE, pp. 332–337 (2009)
21. Min, R., Bhardwaj, M., Cho, S.-H., Shih, E., Sinha, A., Wang, A., Chandrakasan, A.: Low-power wireless sensor networks. In: International Conference on VLSI Design, pp. 205–210 (2001)
22. Min, R., Chandrakasan, A.: Energy-efficient communication for ad hoc wireless sensor networks. In: Signals, Systems and Computers, Conference Record of the Thirty-Fifth Asilomar Conference, vol 1, pp. 139–143 (2001)

23. Anastasi, G., Conti, M., Falchi, A., Gregori, E., Passarella, A.: Performance measurements of motes sensor networks. In: Proceedings of the 7th ACM International Symposium on Modeling, Analysis and Simulation of Wireless and Mobile Systems (MSWiM'04), pp. 174–181 (2004)

24. Chang, J.H., Tassiulas, L.: Energy conserving routing in wireless ad hoc networks. In: Proceedings of the 19th IEEE Conference on Computer Communications (INFOCOM), pp. 22–31 (2000)

Power Optimized Real Time Communication Through the Mobile Sink in WSNs

Ritwik Banerjee and Chandan Kumar Bhattacharyya

Abstract In the recent decades, computer science and engineering has found an emerging research domain in data communication network field, the WSN. WSN applications are generally deployed in fields where uninterrupted supervisions are required. In WSN applications, thousands of energy-constrained sensor nodes are used to sense the data from the deployed environment and transmit the sensed data to the base station (BS). Recent advances in wireless sensor networks lead to rapid development of real-time applications. The requirement of low latency in communication and maximum utilization of node's battery power are becoming more and more important issues in emerging applications especially in fire monitoring, medical care, battle field surveillance, etc. The cluster-based routing protocols can improve the energy life of the sensors and hence can prolong the entire network lifetime but uneven load distribution among the clusters, and static BS may lead to energy hole problem to the entire communication network. In this paper, we present a novel real-time energy-efficient routing protocol in a different way so that real-time communication is achieved by allowing low latency, but it must not incur infinite bounded waiting for the non-real-time regular data communications. This paper also considers mobile sink node to avoid the energy hole problem.

Keywords WSNs · Cluster-based real-time routing protocols · Energy efficiency · Sink mobility · TDMA–CDMA

R. Banerjee (✉)
Department of NIELIT, ICE(I), Techno India Group, Kolkata, India
e-mail: contact.ritwik@gmail.com

C.K. Bhattacharyya
Department of CSE, Techno Saltlake, Kolkata, India
e-mail: ckbtechno@gmail.com

© Springer India 2015
R. Chaki et al. (eds.), *Applied Computation and Security Systems*, Advances in Intelligent
Systems and Computing 304, DOI 10.1007/978-81-322-1985-9_11

153

1 Introduction

Embedded system and wireless communications have led to the invention of new technologies in recent advances—wireless sensor networks. Sensors are mini chips and integrated with the machines and environments, capable of sensing information and delivery of the sensed information. This could provide a great benefit to our society which includes the following: conservation of natural resources, providing homeland security, and improved emergency services. An ideal wireless sensor network is highly scalable, maximizing the battery power, costs very little for installation, reliable enough for information delivery, fast enough for real-time data transfer, and does not require any physical maintenance. Designing the sensor network for a particular application requires the basic knowledge about problem definition, knowledge about node's battery health condition, data transfer rate, bandwidth allocation, admission control algorithm, collision avoidance scheme, etc. As an example, application like temperature monitoring does not require high data rate, but fire-monitoring system, battle field surveillance, and intruder attack require high-rate data transfer [1]. Although quality of service is the primary goal in WSN traffic, low latency in communication and node's energy usage maximization is becoming more and more important in current scenario [2, 3]. Thus, to support real-time data communication, it is needed to provide way for data that are having deadline very close. The real-time data communication should also consider that non-real-time data must not suffer from infinite bounded waiting. In application like intruder attack tracking, surveillance must require the current position of the intruder to be reported to the BS in time. In the same system, there may exist data with different deadlines due to the presence of different requirements. Therefore, developing a real-time routing protocol in WSN should consider timeliness of the communication along with the priority of the packet such that packet deadline missing ratio can be minimized [4, 5]. Designing a real-time routing protocol in WSN is much more challenging because here low-cost sensors are tied with low energy capacity and without having capability of recharging or replacing such that node's health and network lifetime can be maximized [2]. Sensors can transmit data to the BS either by direct communication routing approach or by means of hierarchical routing technique. It is already established that direct communication technique requires much more energy dissipation compared to cluster-based hierarchical routing techniques [2]. The cluster-based routing technique requires the sensors to transmit the data to the sink node though an intermediate node, called cluster head. Here, nodes are clubbed together to form cluster, and a node will be selected within the cluster as cluster head. The cluster head for any cluster is responsible to collect data from the entire sensors, aggregate the collected data, and to transmit the aggregated data to the BS. The energy dissipations for any sensor nodes mainly occurred during data communication period. The maximum energy drainage occurred for the cluster heads, as those special nodes are needed to receive and transmit data for maximum time [6]. Considering cluster-based routing is an effective way to design energy-efficient routing technique, but formation of cluster with irregular distribution of loads may tend to the uneven energy leakage to certain nodes of the network. To prolong the network lifetime, it is required to fix the maximum number of sensor nodes under cluster heads so that no cluster head will come to an end too early [3]. Traditional WSN is composed of stationary sensor node; basically, the immobile BS may incur another serious issue on energy threat—the 'energy hole' problem [6]. The sensors near the BS have to participate in communication on behalf of other sensors and thus will drain their energy very quickly. It results in sudden death for some sensors that are very close to static sink node, and hence, the lifetime of the whole network will be threatened. In this paper, we aim to develop a novel cluster-based real-time routing protocol that will provide a balanced loading between clusters, and to avoid the 'energy hole' problem, we will consider mobility to the BS [6]. We will also consider the real challenge of continuous location updating for the BS without compromising energy efficiency and low latency for real-time and non-real-time data. Hence, for designing the optimal cluster-based routing technique, we will consider the following pivotal issues:

- Designing a cluster-based energy-efficient routing protocol considering even distribution of loads among the clusters.
- Providing low latency for the real-time data communication as well as non-real-time data communication should not be blocked for infinite time.
- Providing mobility to the BS to avoid energy hole problem [6].

The rest of the paper is organized as follows. Section 2 illustrates related works. In Sect. 3, we present the network model used in our algorithm. Section 4 describes the problem definition, whereas in Sect. 5, we describe our proposed protocol. The simulation result is presented in Sect. 6.

2 Related Works

The RAP [4] was designed to support the communication in large-scale network. It follows query-event service and uses triangulation method to determine location of target. It maintains prioritize scheme for real-time data but does not care about energy wastes due to geographic forwarding scheme. The EAQOS [7] protocol finds an optimal solution for real-time as well as non-real-time data by implementing priority queuing model. The SPEED [5] uses beacon exchange mechanism among the sensors for updating the neighbor list for every sensor nodes. SPEED follows stateless non-deterministic geographic forwarding (SNFG) for achieving velocity adjustment and network congestion control. The real-time power awareness [8] develops a power adaptation scheme for assigning velocity to the real-time data packet dynamically after estimating the delays of the packets. Forming clusters and cluster head election based on probability value were the objective of the pioneer clustering algorithms presented by LEACH [2]; later cluster head election technique is modified by taking the node's current energy level into account [9]. A BS-assisted cluster head selection algorithm is presented in [10], where nodes with high sensing density, that is, closer to the entire members, will get higher chance of becoming cluster head. A novel idea for far zone nodes through two-hop communications is presented in [11]. Improvement through a two-tier data disseminations model for the large-scale wireless sensor networks by introducing mobility to the sink nodes is presented in [12]. An adaptive location update for mobile sinks in wireless sensor networks is presented in [13, 14]. Energy-efficient routing schemes for mobile sink in WSNs are established in [15–17].

3 Network Model

Our primary aim is to develop a cluster-based energy-efficient real-time routing protocol. Here, we introduce a novel architecture that ensures well-defined clusters along with real-time data communication protocol. We follow the network model and architecture presented below:

- The sensor nodes are of homogeneous type and deployed in ad hoc manner.
- All the sensor nodes are with uniform initial energy strength, and a unique ID.
- Nodes are eligible to determine their current energy level, channel's signal strength, location information, and ID.
- All the sensor nodes except the BS are immobile. Mobile BS will remain static during the first time cluster formation phase.
- Data aggregations are done at the cluster heads only.
- The communication channels are symmetric.

4 Problem Definition

In general, the cluster-based routing algorithms are segregated into rounds where each round has two phase—setup phase and steady phase. In setup phase, the sensors are generally clubbed together into clusters; from each cluster a special node is selected or elected as an intermediate, called cluster head. In ready phase, the sensors sense and transmit the data to the respective cluster head and the cluster head sends the aggregated data to the BS once in a round. It is clear that the energy dissipation occurred to the cluster heads only as those nodes are needed to turn on the radio to receive and transmit signals. The energy dissipation to the cluster head is directly proportionate to the number of cluster members related to it. If the cluster formation does not scarce about evenly distribution of the loads among the clusters, the cluster head from heavily loaded cluster will losses its energy strength very fast in compare to other cluster heads. Although cluster-based routing techniques are considered as much more energy efficient than geographic packet forwarding techniques, the cluster formation at every round creates an overhead and dissipates huge amount of energy during execution of setup phase. The steady phase in every round, where the real intercluster communication takes place, is scheduled by TDMA approach to avoid packet collision and easy channel access purpose. Here, every node will have to wait for its turn to transmit data to the cluster head; this approach is fine for non-real-time data, but in case of some applications like fire-monitoring system, battle field surveillance, and intruder attack, the out-of-time data are not only irrelevant but may also lead to an adverse situation to the environment. Thus, to support real-time data communication, it is needed to provide immediate way for data that are having deadline very close. The real-time data communication should also consider that non-real-time data must not suffer from infinite bounded waiting. Since all the communications are ultimately headed toward the BS, maximum data packet communications are executed around the sink node. The stationary sink node for any WSN may incur energy hole, another serious problem of heavy energy leakage around the fixed BS and nodes from that particular portion of the network will die very early; hence, the entire network lifetime is threatened. According to the discussion above, we summarize the problem definitions as below:

- Clusters with uneven distribution of loads may destroy the balance of the entire network.
- The static sink nodes create 'energy hole' problem around the BS.
- Real-time data transmitted within time encourage heavy traffic for non-real-time data packet.

5 Proposed Protocol Architecture

The proposed protocol has been divided into rounds where each round is segregated into different phases: the primary setup phase, the regular setup phase, and the steady phase. Our proposed protocol is headed to design an energy-efficient routing algorithm by considering evenly loaded distribution among the clusters. We propose a unique BS-assisted cluster formation technique that is based on very simple architecture but is very efficient to distribute the loads among the clusters in an even manner. The proposed protocol is also aimed to reduce the total number of data communications and reconstruction of clusters at every round. Our proposed protocol ensures a better solution for the real-time as well as for non-real-time data communications. We employed a novel scheme so that non-real-time regular data communication will enjoy a separate queue with traditional TDMA approach for providing high quality of service during data delivery, whereas the real-time data communication uses a different queue with CDMA approach to support successful delivery of data with the minimum time delay.

The CDMA approach was conceived several decades ago, but current advances in electronic technology have finally implemented it successfully. CDMA offers data transmissions with different codes so that multiple sources can transmit their own data simultaneously through the same link at any moment. We propose that at every node in network, there will be one packet qualifier, two different queues, and two different channelization protocols for the real-time and non-real-time data communications. When a data packet arrives at any node, the packet qualifier at that node will check the type of the incoming packet, if it has an early deadline—means real-time traffic—the packet will be transferred to the real-time traffic queue, and hence, the CDMA approach is taken for that packet to deliver it immediately. On the other hand, the non-real-time data packet is taken into different non-real-time queue (NRTQ) and is scheduled for transmission through TDMA time-sharing technique. The primary setup phase executes at the first round and only when the reconstruction of cluster is needed during the entire network lifetime. In other round, the steady phase is followed by the regular setup phase.

5.1 Primary Setup Phase for Cluster Formation, CH Selection, and Schedule Creation

Cluster Formation After being deployed randomly, the nodes are used to create the initial clusters at this phase. Here, we follow the traditional LEACH [2] algorithm to form the clusters; each node uses a probabilistic approach to elect itself as a cluster head at the initial round only. Initially, each node will calculate its priority value for becoming intermediate cluster head by using the following Eq. 1.

$$T(n) = [P/\{1/P \times (r\%1/P)\}] \times [E_n, r] \quad \text{if} \quad n \in G$$
$$\text{Or,} \tag{1}$$
$$T(n) = 0 \quad \text{if} \quad n \notin G$$

where 'n' is the number of nodes inside the network, 'r' is the round number (initial value is 0), and 'P' is the desired percentage of CH for the network. '$E_{n,r}$' is the residual energy level of node 'n' at round 'r'. 'G' is the set of all alive nodes. After calculating the priority value '$T(n)$' for any node, 'n' will check whether this value is greater than a randomly generated threshold value (ranges between 0 and 1) or not. If the node's priority value is greater than the random value, then the node itself will declare as an intermediate CH for that initial setup phase otherwise will be waiting for the CH announcement by the other nodes. The basic idea is that node with higher energy level will be having higher chance of becoming intermediate CH for that primary setup phase. This CH announcement packet contains the location information of the CH node too. The non-CH member node after receiving one or multiple CH announcement selects only the closest one as its intermediate CH and sends its ID and location information to the CH. Each intermediate CH node completes its work by transferring location information (in Cartesian coordinate format) and the current energy status of the entire member nodes from its cluster in an aggregated format to the BS.

After receiving those data from the intermediate CHs, the BS converts those locations to polar coordinate form (r, θ). The BS then divides the total area into subclusters (often called as 'zone') based on the desired percentage of cluster heads, say 'P' with the help of following equations. 'MIN' is the minimum 'weight' per zone, and 'MAX' is considered as the maximum weight per zone [3]. Minimum number of zones,

$$Z_n = P \times 100 \qquad (2)$$

If we consider $P = 0.05$ (means 5 %), then $Z_n = 0.05 * 100 = 5$. Thus, initial degree range per zone

$$D = (360/Z_n)^\circ = 72^\circ \qquad (3)$$

The total area is now equally divided with respect to range D degree, starting from 0 to 360 ° polar coordinates, in anti clockwise direction. Now, BS will find the number of sensors lying per zone, called weight of the zones. If in any situation, it is found that any zone is having less than MIN number of nodes, then that zone is called 'weak zone' and will be merged with its predecessor or successor zone, located in anticlockwise direction. If for any zone, it is found that the weight is greater than the MAX value, and then the zone is to be partitioned again into sub zones, starting from the initial degree of the zone to until MAX number of nodes are found or half of the Zone's weight is met or which appears earlier. The process will be continued until all zones obtain weight between the ranges [MIN, MAX]. Figure 1 is used to present cluster formation through our proposed protocol by considering even load distribution.

Cluster Heads Selection and Schedule Creation After distributing the loads evenly among the clusters, the BS selects two special nodes from each cluster: a real-time cluster head (RTCH) and a non-RTCH (NRTCH). The RTCH is selected through fuzzy implementation by considering parameters like node's current energy level and nodes distance to the BS as well as the signal strength of the sensors, received by the BS itself so that not only distance to the BS but also void zone in network or high traffic congestion can be avoided during RTCH selection phase [18], whereas each NRTCH is selected by the same way but by using an additional parameter—node's distance to the intercluster members. Parameters are chosen so because in case of real-time applications (which occur very often), achieving the lowest latency is the primary goal, whereas in case of non-real-time applications (that is regular), the frequent data communications require optimization of the energy consumption by minimizing the intercluster data communications. Now, the BS sends a traditional TDMA schedule for each node within the cluster along with id of the NRTCH node for that particular cluster, so that the non-real-time regular communication can be continued without any interruption through well-defined non-overlapping TDMA schedule. Every node from each cluster will enjoy some

Fig. 1 The nodes are evenly distributed among the well-defined clusters (*zones*); *red-spotted* nodes are elected as primary cluster heads at the first round only. The MIN and MAX are taken as 10 and 25 for each cluster (*zone*)

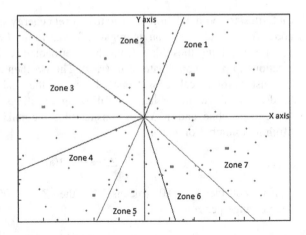

distinct TDMA schedule slots for transmitting data to the NRTCH node during regular data transfer process. Nodes are eligible to transmit data only at its scheduled transmitting slots. For real-time data communication, the routing between cluster member and the respective RTCH will be managed by using CDMA approach only. We consider the real-time data are to be transmitted with minimal latency, and hence, the channel access delay should be optimized. Only the CDMA offers data communications from multiple sources simultaneously through the same link at any moment by using different codes. The necessary direct sequence of the code generator is obtained through traditional 'Walsh Matrices.' For an example, consider four nodes 1, 2, 3, and 4 are connected to the same channel. The data from each station are d1, d2, d3, and d4. The codes assigned to the stations are c1, c2, c3, and c4. The codes are having two properties: (a) multiplication between two different code results 0 and (b) multiplication to self-code will result in 4, the number of nodes. During transmission of data packets, each station at any moment at first multiplies its data with its code and puts on common channel. The data that go on the channel are the sum of all the products from all intended nodes and is of the form (d1.c1 + d2.c2 + d3.c3 + d4.c4). To receive the data from sender, the intended receiver will multiply the integrated data by the sender's code. Suppose to receive data from node 1, receiver multiplies the data with c1 (code for node 1). The resultant data will become (d1.c1 + d2.c2 + d3.c3 + d4.c4) .c1 = (d1.c1.c1 + d2.c2.c1 + d3.c3.c1 + d4.c4.c1) = (4d1 + 0 + 0 + 0) = 4d1.   Dividing   the result by 4 (number of nodes), the receiver can hear to the exact data provided by node 1. Thus, in this way, multiple nodes can access channel simultaneously without suffering from collision. The basic goal is that any real-time data can be sent to the BS through the intermediate RTCH node at any moment with the minimal latency, and this transmission is independent of the transmitting slots defined by TDMA schedule for the regular and non-real-time data communications. Hence, the real-time data and non-real-time data communications are independent of each other and can be initiated separately without suppressing others.

5.2 Regular Setup Phase for Selecting CHs at Non-Initial Round

At the initial round, when the primary setup phase is invoked, the BS seemed to be static, but after the cluster is formed, the only BS is considered as mobile. The mobile BS regularly updates its location information to the NRTCH and RTCH nodes at every round, and when the distance of the BS to the CHs exceeds a predefined threshold value (the one-hop distance of the sensors), the BS selects new RTCH and NRTCH nodes from each cluster and informs the entire cluster members including the old RTCH and NRTCH nodes about the cluster heads. Otherwise after successful completion of each round the new RTCH, NRTCH, TDMA schedule, and CDMA schedule is fixed and conveyed to the entire network. The selection criteria for RTCH and NRTCH are same as described in 'Primary Setup Phase,' the RTCH is selected by considering parameters like node's current energy level and node's distance to the BS, where as each NRTCH is selected by same way but using parameters like current energy level of the nodes and node's distance to the intercluster members, and node's distance to the BS. Both CHSs are selected through well-known 'Mamdani' fuzzy model.

5.3 Steady Phase for Real- and Non-Real-Time Communication

We consider each sensor node in the network has a qualifier to distinguish real-time (RT) and non-real-time data packets. Every data packet is considered to assign a deadline time value before which the packet must be transmitted to the intended recipients. The qualifier always keeps updates about the average end-to-end delay for packet delivery to the BS through NRTCH and RTCH nodes. Every node in network maintains two different queues for real-time and non-real-time communications. When a new packet arrives, the qualifier checks the deadline time of the packet. If the deadline is satisfied by the average end-to-end delay time through NRTCH, the packet is loaded into NRTQ; otherwise, the qualifier puts it to the real-time queue (RTQ). That is, the non-real-time regular data packets are sensed, queued, and transmitted to BS through the NRTCH node from the non-real-time packet queue, whereas real-time data packets are sensed, queued, and transmitted to the BS through the intermediate RTCH node from the real-time packet queue. The non-real-time communication is scheduled by non-overlapping TDMA slots for each cluster members where each member can transmit data at its particular scheduled slots. The real-time data packets are transmitted to the BS through prefixed CDMA codes to provide immediate data transmission scheme. The idea is to provide minimal latency to the real-time data communication and to avoid starvation for the non-real-time data communications. We follow CSMA/CA protocol during intercluster communication period to avoid collisions.

5.4 Algorithm and Pseudo Codes (Partial)

Algorithm for Steady Phase // Assumptions.
/* N is the node itself. P is any packet arrives at node
N. Td is the deadline for packet P. Tavg_delay is the
average end to end delay through NRTCH, computed at node
N. NRTQ is the non-real time queue RTQ is the real time
queue */
// Pseudo Code
For each packet 'P', at any node 'N'
Begin
Find deadline for 'P', say 'Td'
If (Td > Tavg_delay)
Then,
Add 'P' to the NRTQ
Else
Add 'P' to the RTQ
End If
End For
Algorithm for NRT Schedule // Assumptions.
/* N is the node itself. P is any packet arrives at node
N. Tslot_N is the transmitting slot defined by the TDMA
for node N. Tslot_current is the current slot presented
by the TDMA scheduler. Send_Unicast() is the method of
sending packet to in Unicast way. Wait() is the method of
waiting for the next slots*/
// Pseudo Code
For each packet 'P', from front of the NRT Queue, at node
'N'
Begin
If (Tslot_N == Tslot_current)
Then
Send_Unicast (P, N, RTCH, Td)
Else
Wait()
End If
End For
Algorithm for RT Schedule.
/* N is the node itself. P is any packet arrives at node
N. PD is the data part from the packet P. CN is the CDMA
chip code for node N. Send_Unicast() is the communication
to the intended only, to save energy. Wait() is the
method of waiting for the next slots*/
// Pseudo Code
For each packet 'P', from front of the NRT Queue, at node
'N'
PD = PD X CN
Send_Unicast(P, N, RTCH, Td)
End For

6 Performance Analysis and Simulation

We evaluate the performance of our proposed algorithm through simulation experiments; we have implemented the simulator through MATLAB environment. The performance analysis takes into account two vital performance metrics: (1) energy efficiency of the protocol and (2) traffic delay for the real-time communications. We have compared our algorithm with the renowned stateless real-time protocol SPEED. The initial parameters and considerations used for simulation purpose are followed from the traditional SPEED protocol and are described in Table 1.

6.1 Energy Efficiency of the Protocol

Our proposed protocol clubbed the nodes into well-distributed clusters where nodes can transmit sensed data to the previously selected cluster heads only; the packet collision and network congestion are achieved through CDMA and TDMA processes. We improvise the energy requirements efficiently by optimizing the communications by avoiding the beacon exchange phase which can save up to '$N *$ ET $x(k, d) + N*(N-1) * ERx(k))$' Joule /round, where '$P$' is the desired percentage of CH, $ETx(k, d)$ is the transmitting energy required for 'k' bit packet to traverse 'd' unit distance, '$ERx(k)$' is the required energy for receiving a 'k' bit packet, and 'N' is the total number of nodes within the network. The reconstruction of cluster is not done at every round, CHs selection, and schedule creation is done by the assistance of BS to prolong the node's energy life. We consider zero mobility to the BS during the cluster formation phases. Although an extra 'n' bit of

Table 1 Simulation parameters

Parameter name	Parameter value
Radio model—transmitting energy	Eelec * k + Eamp * k * $d2$; d is the distance
Eelec and Eamp	50 nJ/bit, and 100 pJ/bit/m^2
Radio model—receiving energy	Eelec * k; k is packet size
Network bandwidth	200 Kb/s
Packet size	50 bytes
Deployment area	100 m × 100 m
Node placement	Random
Radio range	60 m
Node number	100
Node initial energy	1 J
Base station mobility	0–3 m/s

Fig. 2 The entire energy of the network (100 J) is completely dissipated after 1,042 rounds for SPEED (*green line*), after 1,456 rounds for the proposed protocol (*blue line*). Hence, life is prolonged by almost 40 %

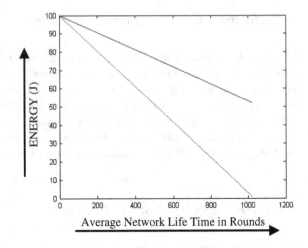

ENERGY (J)

Average Network Life Time in Rounds

'Walsh Code' field is considered in data packet to form different code sequences through 'Walsh matrix of order n', in order to support CDMA data communication from the cluster members to the corresponding RTCH node, here 'n' is the number of nodes within any cluster. The RTCH selection process requires the knowledge about the signal strengths of the entire sensors. The application requires an extra 'n'-bit field, where 2^n is the number of distinct sensors in the network. In our simulation for 100 sensors, we require an extra '7'-bit field ($2^7 > 100$) for every sensor to transmit their signal strengths. This implementation requires an extra 0.35 % of energy dissipation compared to traditional SPEED protocol. The simulation result is shown in Fig. 2.

6.2 Traffic Delay Comparison for Real-Time Communication

Network traffic and packet delay for real-time data packets are measured through time domain. The average delay in real-time routing is determined by calculating the end-to-end time delays and is measured from the entire network with different data packet arrival rates for the nodes. We propose the non-real-time data transmission is scheduled and regularized by the TDMA approach; only the simulation outcome for real-time traffic is shown here. The simulation is done by considering different packet incoming rates—10, 20, 30, 40, 50, 60, 70, 80, 90, and 100 packets/s. The successive delays are measured in seconds. The comparison is done with well-known SPEED protocol. The results are shown in Fig. 3, and it indicates the achievement of minimized latency by implementing CDMA-based immediate data delivery model.

Fig. 3 Delay measurement in real-time traffic, clearly from the figure delay increases as the S packet rate hikes, but our proposed protocol can avoid delay consistently at about 40 % compared to SPEED

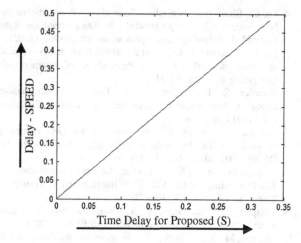

7 Conclusion

In this paper, we concentrate on energy-efficient tracking, dedicated to conserve whole network energy as well as maintain high rate of end-to-end delivery of the packet within the prefixed deadline. We also ensure no additional delay for the non-real-time communications by separating real- and non-real-time traffic during medium access. In order to save energy, we propose minimum computational overhead to the sensor nodes as the cluster formations; RTCH and NRTCH selections and schedule creation are done by BS itself. We propose an immediate CDMA-based intercluster data transmission policy even for higher frequency of real-time data communications. We simulate, analyze, and compare our proposed protocol with the traditional real-time SPEED protocol. The simulation result shows almost 40 % prolonged lifetime along with 40 % less traffic delay. We consider mobile BS to avoid energy hole problem; however, energy-efficient real-time routing by considering mobile sensors is beyond the scope of this paper and considered as the future work.

References

1. Banerjee, R., Bhattacharyya, C.K.: An analysis on soft real time routing protocols in WSN. In: International Conference on TEICC 2012, Bikaner (2012)
2. Heinzelman, W.R., Balakrishnan, H., Chandrakasan, A.: Energy efficient communication protocol for wireless micro sensor networks. In: Proceedings of 33rd Hawaii International Conference on System Science (2000)
3. Banerjee, R., Bhattacharyya, C.K.: Cluster based routing algorithm with evenly load distribution for large scale networks. In: IEEE International Conference, ICCCI 2014, Coimbatore, India (2014, in press)

4. Lu, C., Blum, B., Abdelzaher, T., et al.: RAP: a real-time communication architecture for large-scale wireless sensor networks. In: The Proceedings of the Eighth IEEE Real-Time and Embedded Technology and Applications Symposium (2002)
5. He, T., Stankovic, J., Lu, C., et al.: SPEED: a stateless protocol for real-time communication in sensor networks. In: The Proceedings of International Conference on Distributed Computing Systems, RI, May 2003
6. Banerjee, R., Bhattacharyya, C.K.: Energy efficient routing and bypassing energy-hole through mobile sink in WSN. In: IEEE International Conference, ICCCI 2014, Coimbatore, India (2014, in press)
7. Akkaya, K., Younis, M.: An energy-aware QoS routing protocol for wireless sensor networks. In: The Proceedings of the IEEE Workshop on Mobile and Wireless Networks (MWN2003), Rhode Island, May 2003
8. Chipara, O., He, Z., Xing, G., et al.: Real-time power-aware routing in sensor networks. In: The Proceedings of the 14th IEEE International Workshop on Quality of Service (IWQoS 2006), New Haven, CT, June 2006
9. Tillapart, P., Thammarojsakul, S., Thumathawatworn, T., Saniprabhob, P.: An approach to hybrid clustering and routing in WSNs. IEEEAC (2005)
10. Dhanraj, M., Ram Murthy, S.: On achieving maximum network life time through optimal placement of cluster heads in WSNs. In: Proceedings of IEEE Conference on Communication. IEEE (2007)
11. Katiyar, V., Chand, N., Goutam, G.C., Kumar, A.: Improvement in LEACH protocol for large scale wireless sensor networks. In: Proceedings of IEEE International Conference ICETECT (2011)
12. Ye, F., Luo, H., Cheng, J., Lu, S., Zhang, L.: A two-tier data dissemination model for large-scale wireless sensor networks. In: Proceedings of ACM MobiCom (2002)
13. Nguyan, L.T., Detago, X., Beuran, R., Shinoda, Y.: An energy efficient routing scheme for mobile WSNs. IEEE International Conference (2008)
14. Nasser, N., Yatama, A., Saleh, K.: Mobility and routing in WSN. IEEE CCECE 2011, Canada (2011)
15. Kweon, K., Ghim, H., Hong, J., Yoon, H.: Grid based energy efficient routing from multiple mobile sinks in WSNs. IEEE International Conference
16. Kim, T.H., Adeli, H.: Dynamic routing for mitigating the energy hole based on heuristic mobile sink in WSNs. AST/UCMA/ISA/CAN 2010, LNCS. 6059, Heidelbarg (2010)
17. Wang, J., Yin, Y., Kim, J., Lee, S.: A Mobile Sink Based Energy Efficient Clustering Algorithm for WSNs. IEEE-CIT, Chengadu (2012)
18. Banerjee, R., Bhattacharyya, C.K.: Energy efficient optimization in the LEACH architecture. In: IEEE International Conference, AICERA ICMICR 2013, Kanjirapally, India (2013)

Performance of Incremental Redundancy-Based Data Transmission in Randomly Deployed Wireless Sensor Network

Mousam Chatterjee, Arnab Nandi and Banani Basu

Abstract Energy-level performance of Incremental Redundancy (IR)-based Hybrid Automatic Repeat reQuest (HARQ) scheme using punctured convolution code is evaluated for randomly deployed wireless sensor network (WSN) in the presence of multipath Rician fading. Transmission based on HARQ and optimal power are two different promising approaches for reducing energy consumption in an energy constrained WSN. Optimal transmit power is the minimum power required to sustain the network connectivity while maintaining a predefined maximum tolerable BER threshold in a multihop route. In the present work, energy-level performance of HARQ scheme is compared with that of optimal transmit power-based coded and uncoded schemes for a random WSN. Further, energy consumption for an arbitrary fixed power-based coded scheme is also compared. In a random network, an intermediate node in the route selects the nearest node within a sector of angle (θ) toward the direction of the destination as the next hop. Effects of fading, node density, and search angle on selection of optimum power, energy consumption of optimum power-based scheme, and IR scheme are investigated. Effects of code rate and bit rate on energy consumption, route BER, and optimum power selection in case of optimum power-based coded scheme are indicated.

Keywords Incremental redundancy · Hybrid automatic repeat request · Punctured convolution code · Wireless sensor network · Random network

M. Chatterjee (✉)
B. P. Poddar Institute of Management and Technology, Kolkata, India
e-mail: mousam.chatterjee@gmail.com

A. Nandi · B. Basu
Dr. B. C. Roy Engineering College, Durgapur, India
e-mail: nandi_arnab@yahoo.co.in

B. Basu
e-mail: basu_banani@yahoo.in

© Springer India 2015
R. Chaki et al. (eds.), *Applied Computation and Security Systems*, Advances in Intelligent Systems and Computing 304, DOI 10.1007/978-81-322-1985-9_12

1 Introduction

In wireless sensor network (WSN), the data transmitted from the sensor nodes are vulnerable to errors induced by noisy channels and other factors. Further, due to the severe energy constraint, it is not viable to raise the signal power of the transmitted signal in WSNs [1]. Moreover, high transmission power introduces excessive amount of inter-node interference. If the transmit power is too small, the network might be disconnected due to single or multiple link failure(s). Hence, an alternative way is to use the error control schemes to enhance the packet transmission reliability. ARQ and FEC are the key error control strategies used in WSNs [2]. However, usage of ARQ is limited for sensor networks due to the additional retransmission energy cost and overhead [1]. Hence, it is necessary to provide a proper error control scheme to improve the BER performance while reducing energy consumption. Due to the stringent energy constraint in sensor networks, it is very important to employ energy efficient error control scheme. Incremental redundancy (IR)-based techniques are gaining interest in order to achieve data transmission with low error probability in packet oriented wireless links. Different network conditions such as multipath fading have significant impact on optimal transmit power, energy consumption, and BER performance. So it is important to investigate the impact of multipath Rician fading on network performances, such as optimal transmit power, energy expenditure. However, most of the literature considers regular network architecture without considering fading and shadowing effect in the propagation path. Wireless channels are often accurately modeled as exhibiting selective Rician fading, where there is a strong signal component. Rician fading captures a wide range of fading model. It represents Rayleigh fading when $K = 0$, and no fading when $K \to \infty$, where K is the Rician factor defined as the power ratio of specular to diffused components [3].

In this chapter, performance of IR-based Hybrid Automatic Repeat reQuest (HARQ) error control scheme is evaluated for a randomly deployed WSN in the presence of multipath Rician fading. Incremental redundancy is a HARQ technique in which instead of sending simple repeats of the entire coded packet; additional redundant information bit is incrementally transmitted if the decoding fails on the first attempt. IR scheme uses an infinite ARQ scheme at lowest available code rate if decoding fails with all the incremental bits. Non-systematic convolution codes are used in the present study since efficient algorithms are available [4] for decoding of such punctured code. In our work, a scheme utilizing optimum transmit power is also considered for both convolution-coded and uncoded data and the performance of the schemes is compared to that of IR scheme to present a comprehensive overview. In the optimal power-based scheme, an optimal transmit power corresponding to a particular node density and network condition is exploited for transmitting data. Optimal transmit power is the minimum power required to sustain the network connectivity while maintaining a predetermined maximum tolerable BER threshold in a multihop route [5]. In an ideal scenario, the transmit power of a node should be modified on a link-by-link basis to achieve the maximum possible power savings [6–10]. However, in ad hoc network, performing power control on a link-by-link basis is a complicated and cumbersome task. A straightforward solution in the view of practical implementation is to use a common transmit power for all the nodes. This is very much desirable in inaccessible terrain, where adjustment of the transmit power after deployment is impossible or very much costly. Moreover, the performance disparity, in terms of traffic carrying capacity, between adjusting the power locally and employing a common transmit power is small, especially when the number of nodes is large [11]. Hence, investigation on optimal transmit power under different network condition is important to enhance network lifetime. More precisely our contributions in this chapter are as follows:

1. Evaluation of energy consumption of IR-based packet transmission for a randomly deployed WSN and impact of Rician fading, node density on energy consumption.
2. Optimal transmit power for uncoded and convolution-coded data (with different code rates), different bit rates, and node spatial density are derived for a WSN deployed-based on a random topology.
3. Impacts of severity of multipath Rician fading on optimal transmit power and route BER is studied for such random WSN.
4. In random network, an intermediate node in the route selects the nearest node within a sector of angle (θ) toward the direction of the destination as the next hop. The effects of search angle (θ) on optimal transmit power, energy consumption, and route BER performance are also indicated.
5. Energy consumption of IR-based scheme is compared with optimal power-based scheme (for both using coded and uncoded data) and an arbitrary fixed power-based scheme (with fixed code rate).

2 Related Works

Implementation of incremental redundancy-based HARQ error control scheme in WSN has been widely investigated in recent years. Next, we will review some typical related woks.

Stanojev et al. [12] show that relevant gains in terms of energy efficiency can be achieved by resorting to HARQ protocols rather than conventional time-diversity schemes, where the number of retransmissions is fixed. Rossi et al. [13] presented a reprogramming system for WSNs called SYNAPSE to improve the efficiency of the error recovery phase. SYNAPSE features a HARQ solution where data are encoded prior to transmission and incremental redundancy is used to recover from losses, thus considerably reducing the transmission overhead. In [14], throughput of hybrid automatic retransmission request schemes based on IR using Low Density Parity Check (LDPC) is discussed. Hunter et al. introduced a new method called coded cooperation that integrates channel coding into cooperation diversity

[15, 16]. It was shown that coded cooperation outperforms repetition-based protocols for various scenarios because of its ability to vary its coding rate and thus to adapt to channel conditions [16].

3 Network Model and Problem Description

In this section, we describe the wireless sensor network model under present investigation and the basic assumptions considered in the chapter.

3.1 Network Architecture

We consider a random topology of network. It is assumed that N numbers of nodes are distributed over a region of area 'A' obeying random topology. To avoid edge effects, we assume the network surface to be the surface of a torus with length 2R on each edge. A torus can be formed by connecting the top edge with the bottom edge and the left edge with the right edge of plane [17]. The node spatial density ρ_{sq} is defined as number of nodes per unit area, i.e., $\rho_{sq} = N/A$. Since the positions of nodes in the network are independent and uniformly distributed, it can be shown that the number of nodes in an area has a two-dimensional Poisson distribution [5]. The probability mass function (pmf) of the number of nodes 'N_a' over a surface of area 'a' in the case with random topology is given by [5]

$$P_r(N_a = j) = \frac{(a\rho_{sq})^j}{j!} e^{-a\rho_{sq}} \quad j = 0, 1, 2, \ldots \tag{1}$$

where $a\rho_{sq}$ is the expected number (λ) of nodes present in area 'a'. It is assumed that all sensor nodes are stationary in the present study.

3.2 Routing Protocol

We consider a routing scheme where each intermediate node in a multihop route relays the packets to its nearest neighbor in the direction of the destination as shown in Fig. 1. In particular, we assume that an intermediate node in the route selects the nearest node within a sector of angle 'θ' toward the direction of the destination as the next hop [17]. Let W be a random variable denoting the distance to the nearest neighbor in a two-dimensional Poisson node distribution. It can be shown that, keeping the node spatial density fixed, for large 'N' (i.e., as $N \to \infty$), the CDF of the distance to the nearest neighbor within a sector angle of 'θ' in a torus is [5]:

Fig. 1 Possible multihop
route in a random topology

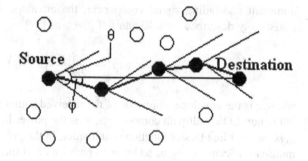

$$F_W^{(\theta)}(w) = 1 \qquad\qquad w > R$$
$$= 1 - e^{-\frac{\rho_{sq}\theta w^2}{2}} \quad 0 \leq w \leq R \qquad (2)$$
$$= 0 \qquad\qquad\quad \text{otherwise.}$$

3.3 MAC Protocol

Here, we consider a simple reservation-based MAC protocol, called REServe-and-GO (RESGO) following [18]. According to this protocol, a source node first reserves intermediate nodes on a route for relaying its packets to the destination. A transmission can begin only after a route is discovered and reserved. If the destination node is busy, it waits for an exponential random back-off time before transmitting or relaying each packet. When the random back-off time expires, node starts transmitting a packet. The random back-off time helps to reduce interference among nodes in the same route and also among nodes in different routes. Thus, as per the present protocol, a node in a route can receive packet or data from only one node at any given time. Throughout this chapter, we assume that the random back-off time is exponential with mean $1/\lambda_t$, where λ_t is the packet transmission rate.

3.4 Channel Model

The major perturbations in wireless transmission are large-scale fading and small-scale fading [19, 20]. Large-scale fading represents the average signal power attenuation or path loss due to motion over large areas. This phenomenon is affected by prominent terrain contours (hills, forests, billboards, clumps of buildings, etc.) between the transmitter and receiver. However, small-scale fading exhibits rapid changes in signal amplitude and phase as a result of small changes (as small as a half-wavelength) in the spatial separation between a receiver and transmitter. If the multiple reflective paths are large in number and there is a dominant non-fading signal component, the envelope of the received signal is statistically described by a Rician pdf given as [19]

$$p_z(z) = z/\sigma^2 \exp\left[\frac{-(z^2 + v^2)}{2\sigma^2}\right] I_0\left(\frac{zv}{\sigma^2}\right) \quad z \geq 0 \tag{3}$$

where z is the envelope amplitude of the received signal, $2\sigma^2$ is the average power in the non-LOS multipath components, v^2 is the power in the LOS component, and I_0 is the modified Bessel function of 0th order. In the present work, we consider the multipath Rician fading in addition to path loss and thermal noise.

3.5 Transmission Schemes

In this chapter, performance of three different transmission schemes in random WSN are evaluated and compared. Here, performance of an arbitrary transmit power-based scheme, optimal power-based scheme, and IR-based scheme are compared. In arbitrary power-based scheme, convolution-coded message is used, while in case of optimal power-based scheme, both convolution-coded and uncoded messages are considered. However, IR-based scheme uses punctured convolution code to evaluate and compare the performances.

3.6 Connectivity

The number of hops depends on the number of nodes available in the network, search angle 'θ', and distance between the source and destination. However, in our analysis, we considered average number of hops to evaluate the performance WSN. The average number of hops also depends on the search angle (θ) and number of nodes in the network. For instance, a route will consist of many hops if each hop length is short. Hop length depends on the value of search angle (θ). The average number of hops on a route is estimated as [17]

$$\bar{n}_{\mathrm{rndm}} \cong \frac{\sqrt{N}\sqrt{\theta^3}\left[\sqrt{2} + \ln\left(1 + \sqrt{2}\right)\right]}{6\sqrt{2\pi} \sin\left(\frac{\theta}{2}\right)}. \tag{4}$$

3.7 BER at the End of a Multihop Route

The received signal at the receiver is the sum of three components (1) the intended signal from the transmitter, (2) interfering signals from other active nodes, and (3) thermal noise. Interfering signals from other nodes are independent identically distributed. Hence, from central limit theorem, it is Gaussian in nature and treated as an additive noise process independent of thermal noise. The received signal in the receiving node Y during each bit period can be expressed as [5]

$$Y = h_s S_{rcv} + \sum_{j=1}^{N-2} h_j s_j + n_{thermal} \tag{5}$$

where h_s and h_j are the channel coefficients of the intended and interference signal, respectively, S_{rcv} is the desired signal in the receiving antenna considering only path loss, s_j is the interference from the other nodes, and $n_{thermal}$ is the thermal noise signal. We also assume that interference from other active nodes (i.e., s_j) undergo similar Rician fading as the desired signal.

Assuming binary phase-shift keying (BPSK) modulation, there can be two cases for the amplitude of the S_{rcv}

$$S_{rcv} = \begin{cases} \sqrt{\frac{P_{rcv}}{R_{bit}}} = \sqrt{E_{bit}} \; ; \text{ for a} + 1 \text{ transmission} \\ -\sqrt{\frac{P_{rcv}}{R_{bit}}} = -\sqrt{E_{bit}} \; ; \text{ for a} - 1 \text{ transmission} \end{cases} \tag{6}$$

where P_{rcv} is the power received at the receiving end, R_{bit} is the bit rate, and E_{bit} is the bit energy of the received signal considering only path loss. P_{rcv} is given by Frii's transmission formulae [19]

$$P_{rcv} = \frac{P_t G_t G_r c^2}{(4\pi)^2 f_c^2 w^\gamma} \tag{7}$$

where P_t is the transmit power, G_t is the transmitting antenna gain, G_r is the receiving antenna gain, f_c is the carrier frequency, γ is the path loss exponent, and c is the velocity of light. Here, we considered omni-directional ($G_t = G_r = 1$) antennas at the transmitter and receiver. The carrier frequency is in the unlicensed ISM band (2.4 GHz).

For each interfering node j, the amplitude of the interfering signal can be of three types with different probability [17]

$$s_j = \begin{cases} \sqrt{\frac{P_{intj}}{R_{bit}}} \text{ with probability } \frac{1}{2} P_{trans} \text{ for} + 1 \text{ transmission} \\ -\sqrt{\frac{P_{intj}}{R_{bit}}} \text{ with probability } \frac{1}{2} P_{trans} \text{ for} + 1 \text{ transmission} \\ 0 \quad \text{with probability } (1 - P_{trans}) \text{ for no transmission} \end{cases} \tag{8}$$

where P_{trans} is the transmission probability of interfering nodes and P_{intj} is the interference power received from node j. The probability that an interfering node will transmit and cause interference depends on the MAC protocol used. Considering the RESGO MAC protocol and assuming that each node transmits packets with length L_T, the interference probability is equal to the probability that an interfering node transmits during the vulnerable interval of duration L_T/R_{bit}, where R_{bit} is the bit rate. This probability can be written as [18]

$$p_{trans} = 1 - e^{-\frac{\lambda_t L_T}{R_{bit}}} \tag{9}$$

Size of the interference vector $\vec{S}_{int}$ increases as the number of nodes increases in the network. The vector $\vec{S}_{int}$ is defined as: $\vec{S}_{int} = \{s_j\}_{j=1,2,\ldots,(N-2)} = \{s_1, s_2, \ldots, s_{N-2}\}$, where s_j (as given in Eq. (8)) is the amplitude of the signal received at the receiver from an interfering node j.

The thermal noise signal can be written as [19]

$$n_{thermal} = \sqrt{FkT_0B} \tag{10}$$

where F is the noise figure, $k = 1.38 \times 10^{-23}$ J/K is the Boltzmann's constant, T_0 is the room temperature, and B is the transmission bandwidth.

Assuming that the threshold for bit detection is placed at 0, the bit error probability can be written as [5]

$$P(\text{bit error}) = \sum_{\vec{S}_{int}} P\left\{ h_s S_{rcv} + \sum_{j=1}^{N-2} h_j s_j + n_{thermal} < 0 \middle| \vec{S}_{int} \right\} P\left\{\vec{S}_{int}\right\}$$

$$= \sum_{\vec{S}_{int}} Q\left(\frac{h_s S_{rcv} + \sum_{j=1}^{N-2} h_j s_j}{\sigma} \right) P\left\{\vec{S}_{int}\right\} \tag{11}$$

where $\sigma = \sqrt{FKT_0/2}$. Assuming that a bit detected erroneously at the end of a link is not corrected in successive links, the BER at the end of a multihop route with $\bar{n}_{rndm}$ number of hops is denoted as BER_{route}. So the BER_{route} can be expressed as [17],

$$BER_{route} = 1 - \prod_{i=1}^{\bar{n}_{rndm}} (1 - BER_{hop\_i}) \tag{12}$$

where BER_{hop_i} is the BER of the ith link of a route. We simulate the BER for link and route following the guiding equations as presented in Eqs. (11) and (12).

3.8 Optimal Common Transmit Power for Random Networks

In a network with random topology, the common transmit power used by each node should be large enough so that the BER at the end of a multihop route with an average number of hops $\bar{n}_{rndm}$, given by Eq. (4), is lower than the maximum tolerable value, denoted as BER_{th}. Thus, the transmit power is chosen in such a way so that the following inequality must be satisfied:

$$BER_{route} \leq BER_{th} \tag{13}$$

We determine the optimal transmit power with the help of simulation test bed developed by us. Transmission power of the node is increased gradually (keeping other parameters fixed) in step starting from a very low power till desired BER_{th} is satisfied. The minimum transmit power satisfying desired BER_{th} is the optimal common transmit power. This optimal transmit power also depends on the coding scheme used. It is obvious that optimal transmit power for coded transmission is less than that of uncoded scheme.

3.9 Algorithm for IR Scheme

The following steps are used for implementing IR scheme in random WSN:

1. Convolution-coded data with code rate of $1/2$ is generated.
2. Every third bit of the coded data is punctured (to obtain effective code rate of $3/4$) and transmitted.
3. If decoding fails, every 6th bit is transmitted (to obtain a effective code rate of $3/5$) and combined with previous transmission for decoding.
4. If again decoding fails, remaining bits are transmitted (to obtain lowest available code rate of $1/2$) and decoded.
5. For further failure of decoding, infinite ARQ scheme is utilized at lowest available code rate.

3.10 Energy Model

Next, we derive the energy spent in successfully transmitting a data packet considering incremental redundancy (IR)-based HARQ error control scheme between a pair of source and destination nodes via multihop route. In our study, non-systematic punctured convolution codes are used as part of IR.

It is assumed that each packet (L_T) consists of header (L_h) and convolution-coded message (L_c). L_c depends on code rate and given as L_m/C_r; where L_m and C_r are the uncoded message length and code rate used. So the energy required to transmit a single packet is

$$E_t = \frac{P_t L_T}{R_{bit}} \tag{14}$$

Here, it is assumed that 75 % of the transmit energy is required to receive a packet [4]. So energy required to communicate, i.e., transmit and receive a single packet is given by

$$E_{\text{packet}} = \frac{P_t(L_T + l_{\text{ack}})}{R_{\text{bit}}} \times 1.75 + (E_e + E_d)L_m \tag{15}$$

where E_e, E_d, and l_{ack} are the encoding energy per useful bit, decoding energy per useful bit, and acknowledge frame length, respectively. In general, for convolution codes, the energy required to encode data is negligible. However, performing decoding on an Intel® StrongARM® SA-1110 processor [21, 22] is energy-intensive. It is seen that the energy consumption in decoding (E_d) seems to be independent of the coding rate [21]. This is reasonable since the rate only affects the number of bits sent over the transmission. A lower rate code does not necessarily increase the computational energy since the number of states in the Viterbi decoder is unaffected [21]. However, the energy consumption scales exponentially with the constraint length [23]. This is expected since the number of states in the trellis increases exponentially with constraint length [24–26]. Thus, the energy required to communicate a single coded packet of size L_c is:

$$E_{\text{packet}} = \frac{P_t(L_h + L_c + l_{\text{ack}})}{R_{\text{bit}}} \times 1.75 + E_d L_m \tag{16}$$

Average probability of error at packet level at each hop is expressed as [27]

$$\text{PER}_{\text{hop\_i}} = 1 - (1 - \text{BER}_{\text{hop\_i}})^{L_T} \tag{17}$$

The effect of fading and code rate are incorporated in BER_{hopi}. We consider two cases: (1) uncoded data transmission and (2) coded data transmission using convolution code. In case of coded transmission, $\text{BER}_{\text{hop\_i}}$ is the error rate incorporating the coding scheme. Since $\text{BER}_{\text{hop\_i}}$ with FEC is different (likely to be less) than that of uncoded case, optimal power, which depends on BER, will also be different compared to uncoded transmission. The probability of 'n' retransmissions is the product of failure in the $(n-1)$ transmissions and the probability of success at the nth transmission [27]:

$$P_I[n] = (1 - \text{PER}_{\text{hop\_i}})(\text{PER}_{\text{hop\_i}})^{n-1} \tag{18}$$

Average number of retransmissions, assuming an infinite ARQ is given as

$$R_I = \sum_{n=1}^{\infty} P_I[n] \cdot n \tag{19}$$

The energy consumed per packet at each hop is considered as the energy spent in forward transmission of information and reverse transmission for NACK/ACK as in [28]. It is assumes that after first unsuccessful transmission (considering incremental bits), the receiver asks for retransmissions at lowest code rate. The energy consumption in ith hop is given as:

$$
\begin{aligned}
E_{hopi} &= E_{R0} + E_{R1} \cdots + E_{Rt} + E_{dec} + E_{retrans} \\
&= \left\{ \frac{1.75P_t \left(L_h + \frac{L_m}{C_{R0}} + l_{ack} \right)}{R_{bit}} \right\} + \left\{ \frac{1.75P_t \left(L_h + \left(\frac{L_m}{C_{R1}} - \frac{L_m}{C_{R0}} \right) + l_{ack} \right)}{R_{bit}} \right\} \\
&\quad + \left\{ \frac{1.75P_t (L_h + \left(\frac{L_m}{C_{Rt}} - \frac{L_m}{C_{R(t-1)}} \right) + l_{ack})}{R_{bit}} \right\} + (t+1)E_d L_m \\
&\quad + \left\{ \frac{1.75P_t \left(L_h + \frac{L_m}{C_{Rm}} + l_{ack} \right)}{R_{bit}} + E_d L_m \right\} R_I \\
&= \frac{1.75P_t}{R_{bit}} \left\{ (t+1)(L_h + l_{ack}) + \frac{L_m}{C_{Rm}} \right\} + \left\{ \frac{P_t \left(L_h + \frac{L_m}{C_{Rm}} + l_{ack} \right)}{R_{bit}} \times 1.75 + E_d L_m \right\} R_I
\end{aligned}
$$

$$(20)$$

where E_{Rt} is the energy associated with tth transmission of incremental bits. In our study, we considered $t = 2$. E_{dec} is the total decoding energy used to decode the message using incremental bits. $E_{retrans}$ is the energy required to communicate data using infinite ARQ scheme at lowest available code rate (if decoding fails with all the incremental bits). C_{R0} is the code rate used for initial transmission. C_{Rt} is the effective code rate after tth transmissions of incremental bits. Thus, total energy required to communicate a packet at the end of $\bar{n}_{rndm}$ number of hops is given as:

$$
E_{total} = \sum_{i=1}^{\bar{n}_{rndm}} E_{hop\_i} \tag{21}
$$

4 Simulation Model

We now present the simulation model developed in MATLAB to evaluate the performance of IR in WSN. Simulation model is based on the IR algorithm given in Sect. 3.9. Mean value of 20 independent runs (each run of 10^5 bit) in each case is considered for presentation. Table 1 shows the important network parameters used in the simulation study:

- Digital data 1 and 0 with equal probability is generated at base band.
- Digital data passed through convolution encoder with code rate ½.

Table 1 Network parameters used in the simulation

Parameter	Values
Path loss exponent (γ)	2
Number of nodes in the network (N)	1,000
Node spatial density (ρ_{sq})	10^{-8}–10^{-1}
Packet arrival rate at each node (λ_t)	1 pck/s
Career frequency (f_c)	2.4 GHz
Noise figure (F)	6 dB
Room temperature (T_0)	300 k
Transmission power (P_t)	1 mW
Search angle (θ)	$\pi/9$, $\pi/2$, and π
Constraint length of encoder	7
Decoding energy per useful bit (E_d)	1.2×10^{-4} J

- Every third bit of the coded data is punctured and transmitted. Our transmitted signal is +1 or −1 corresponding to data 1 or 0.
- Fading channel coefficients are generated following Rician distribution as in Eq. (3).
- The distance between the transmitter and the receiver (i.e., W) is randomly generated from the hop length distribution given in Eq. (2).
- Next, a number of interfering nodes within the circle of radius 2 W centered at the receiver are generated according to a two-dimensional Poisson distribution with mean '$a\rho_{sq}$'.
- Active interfering nodes are identified using Binomial distributed random variables for node activity.
- Interference from such active interfering nodes is generated assuming interference undergoes similar kind of fading as the desired signal.
- The desired message signal is affected by multipath fading, thermal noise, and interference from other nodes. The signal received by the receiving antenna in destination node is generated following Eq. (5).
- The received signal Y as given in Eq. (5) is then detected considering the threshold level at 0. If the received signal is greater than the threshold level 0, then it is detected as 1. Otherwise, it is detected as 0.
- Each received bit is then compared with the transmitted bits. If there is mismatch an error counter is incremented. Now dividing the error count by the total number of transmitted bits, link BERs are obtained.
- In case of unsuccessful decoding, every 6th bit of the initially coded data is transmitted and decoded.
- If again decoding fails, remaining bits are transmitted and decoded.
- If decoding fails with all the incremental bits, system utilizes infinite ARQ technique at lowest available code rate.
- Transmit power is increased gradually in step starting from a very low power till desired BER_{th} is satisfied. The minimum transmit power satisfying desired

BER_{th} is the optimal common transmit power. Thus, optimal transmit power for various bit rate, search angle (θ), and node spatial density is obtained.

- The energy consumption in IR scheme in random network is evaluated using Eq. (21).

5 Simulation Results

In this section, the impact of several network parameters on energy consumption is discussed to present a comprehensive overview. Simulations are performed using MATLAB®, and results are compared in terms of mean value of 20 independent runs in each case. The simulation parameters are listed in Table 1. Confidence intervals are shown only for Fig. 2. However, it is always at a confidence level of 95 %.

Figure 2 shows route BER as a function of node spatial density for uncoded and convolution-coded data. It is observed that BER_{route} performance improves with increase in node spatial density. However, it is seen that beyond a certain node density, the BER_{route} does not change with further increase in node spatial density and a floor in BER_{route}, as denoted by BER_{floor} appears. The desired signal power as well as the inter-node interference increases with increase in node density. As a result, we obtain the BER_{floor}. This is expected because, increasing node spatial density beyond a certain limit no longer improves the signal-to-noise ratio (SNR), as the interfering nodes also become close enough to the receiver. It is seen that BER_{route} performance improves in case of convolution-coded message as compared to uncoded message. Further route BER performance improves with decrease in convolution code rate.

Fig. 2 Route BER as a function of node spatial density for different code rate in the presence of Rayleigh fading; $R_{bit} = 10$ Mbps, $\theta = \pi/9$

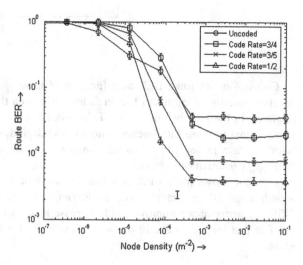

Fig. 3 Route BER as a
function of node spatial
density for different severity
of fading; code rate = ½,
R_{bit} = 10 Mbps, $\theta = \pi/9$

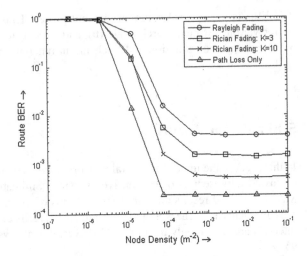

Fig. 4 Route BER as a
function of node spatial
density for different search
angle in the presence of
Rayleigh fading; code
rate = 3/5, R_{bit} = 10 Mbps

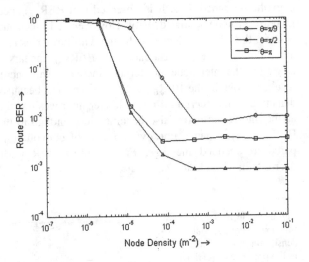

Figure 3 shows route BER as a function of node spatial density for different
level of severity of multipath Rician fading. It is seen that BER_{route} performance
degrades in the presence of fading. This is because in multipath fading environment, signal-to-noise interference ratio (SNIR) degrades. It is also observed that
with increase in severity of fading, i.e., as K factor decreases from 10 to 3,
BER_{route} performance degrades.

Figure 4 shows route BER as a function of node spatial density for different
search angle (θ) in the presence of Rayleigh fading ($K = 0$). It is seen that
BER_{route} performance improves with increase in search angle (θ). This is due to
the fact that for high value of the search angle (θ), the hop length is likely to be
short.

Fig. 5 Optimum transmit
power as a function of bit rate
for different fading severity;
code rate = 3/4, node
density = 10^{-6}, $\theta = \pi/9$

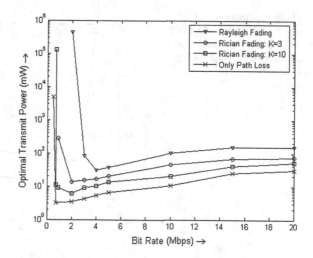

In Fig. 5, we compare the optimal common transmit power in a random net-work as a function of bit rate in multipath Rician fading channel. Optimal common transmit power is the minimum transmit power sufficient to preserve network connectivity while satisfying a predetermined BER threshold (BER_{th}) value at the end of a multihop route. It is seen that optimal transmit power increases as the data rate increases. It is mainly because of the high thermal noise introduced due to high bit rate. It is observed that optimal transmit power required to transmit data in multipath Rician fading channel is higher than the power required in the absence of fading for same data rate. Further optimal transmit power increases with increase in severity of fading. For example, at a bit rate of 15 Mbps, $K = 10$ and $BER_{th} = 10^{-2}$, and the optimal transmit power is 40.5 mW. However, for the same BER_{th} and data rate, the optimal transmit power is increased to 68.5 mW for K value of 3. Further, there is a critical data rate, below which the desired BER_{th} cannot be satisfied for any level of transmit power. The critical bit rate occurs at the point where the BER_{floor} for that particular data rate becomes higher than the desired BER_{th}. It is also seen that critical bit rate increases with increase in severity of fading.

In Fig. 6, we compare the optimal common transmit power in a random net-work as a function of bit rate for several code rates of convolution-coded data. It is seen that optimal transmit power increases as the data rate increases. Further, critical bit rate increases with increase in data rate. For example, critical bit rate is 0.3 Mbps at a code rate of ½, while it increases to 3 Mbps at a code rate of ¾.

Figure 7 shows the optimal common transmit power as a function of node spatial density for different search angle (θ). It also shows the effect of convolution coding on optimal transmit power in a multipath Rayleigh fading channel. It is seen that optimal transmit power increases with decrease in node spatial density for a given bit rate. This is because with decrease in node spatial density at a fixed bit rate, BER_{route} performance degrades. It is seen that optimal transmit power decreases with the use of coded message and as code rate increases optimal

Fig. 6 Optimum transmit power as a function of bit rate for different code rate in the presence of Rayleigh fading; node density $= 10^{-6}$, $\theta = \pi/9$

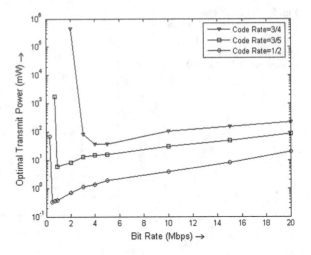

Fig. 7 Optimum transmit as a function of node density for different search angle in the presence of Rayleigh fading; $R_{\text{bit}} = 10$ Mbps

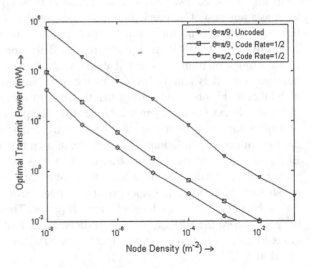

transmit power increases. Further optimal transmit power required to maintain desired BER threshold increases with decrease in search angle (θ).

Figure 8 shows the energy required to successfully deliver a typical message of size 100 bit using IR-based HARQ technique in the presence of Rayleigh and Rician fading. Energy consumption of IR scheme is compared with that of coded scheme which uses an arbitrary fixed transmit power with several code rates. Energy requirement decreases significantly in case of IR-based scheme compared to that of fixed code rate-based scheme. It is seen that energy requirement increases in IR scheme in the presence of fading. Further, in IR scheme energy requirement decreases with increase in search angle (θ). Energy consumption decreases with decrease in code rate of convolution-coded message.

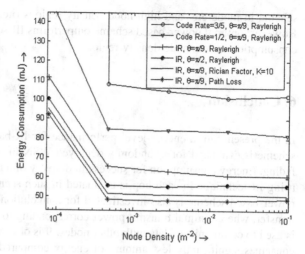

Fig. 8 Energy consumption to communicate 100 bit of data for different code rate in the presence of fading using fixed transmit power of $P_t = 1$ mW; $R_{bit} = 10$ Mbps

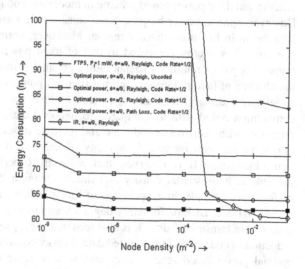

Fig. 9 Energy consumption to communicate 100 bit of data utilizing fixed and optimum power in the presence of fading; $R_{bit} = 10$ Mbps. FTPS is the abbreviated form of fixed transmit power scheme

Figure 9 shows the energy required to successfully deliver a message of size 100 bit using optimal transmit power corresponding to each node spatial density. Energy requirement in optimal power-based scheme increases in the presence of fading. It is seen that use of optimal transmit power-based scheme reduces energy requirement significantly compared to fixed power-based transmission. Further, energy requirement decreases with increases in search angle 'θ'. In case of optimal power-based scheme, energy consumption increases in uncoded transmission of message compared to that of coded transmission. Moreover, it is observed that IR-based scheme consumes less energy than those of optimal power-based scheme (for both coded and uncoded message) as well as arbitrary fixed power-based scheme in moderate- and high-node density regions (i.e., our region of interest). However, optimal power-based scheme outperforms IR scheme in terms of energy consumption in low-node density region.

6 Conclusion

In the present work, energy-level performance for IR-based HARQ error control scheme is evaluated for a randomly deployed WSN in the presence of multipath fading. Energy consumption for successful delivery of a packet via multihop route using IR technique in each hop is evaluated in such scenario. Further, an optimal power-based scheme is also investigated for convolution-coded and uncoded data transfer, where optimal transmit power corresponding to a particular node density is used to communicate data with other nodes. It is observed that IR-based scheme consumes significantly less amount of energy compared to optimal power-based scheme and fixed power-based scheme in moderate- and low-node density regions. However, optimal power-based scheme consumes less energy compared to that of IR scheme in low-node density region. Moreover, optimal power-based scheme consumes less energy compared to that of fixed power-based scheme. Optimal transmit power in multipath Rician fading channel is more as compared to that in the absence of fading. Further, optimal transmit power decreases with increase in node spatial density. It is also seen that optimal transmit power decreases for convolution-coded data as compared to uncoded data. Optimal transmit power decreases with decrease in code rate and increase in search angle (θ). Critical bit rate increases with increase in severity of fading and increase in code rate of transmitted data. It is observed that route BER performance improves with increase in node spatial density and finally attains a floor. Further, route BER performance degrades in the presence of Rician fading and with decrease in search angle (θ). Route BER performance degrades with increase in severity of fading and code rate of transmitted data. It is also seen that energy requirement decreases with reduction in code rate of transmitted data. Energy consumption for both the IR and optimal power-based scheme decreases with increase in search angle 'θ'. The study helps in designing energy efficient randomly deployed WSN.

References

1. Akyildiz, I.F., Weilian, S., Sankarasubramaniam, Y., Cayirci, E.: A survey on sensor networks. IEEE Commun. Mag. **40**(8), 102–114 (2002)
2. Nandi, A., Kundu, S.: Energy level performance of error control schemes in wireless sensor networks. IEEE International Conference on Devices and Communications (ICDeCom 2011), pp. 1–5, 2011
3. Davarian, F.: Fade margin calculation for channels impaired by Rician fading. IEEE Trans. Veh. Technol. **34**(1), 41–44 (1985)

4. Sankarasubramaniam Y., Akyildiz I.F,. Mclaughlin S.W.: Energy efficiency based packet size optimization in wireless sensor networks. Proceedings of the First IEEE International Workshop on Sensor Network Protocols and Applications 2003, pp. 1–8, 2003
5. Panichpapiboon, S., Ferrari, G., Tonguz, O.K.: Optimal Transmit Power in Wireless Sensor Networks. IEEE Trans. Mob. Comput. 5(10), 1432–1447 (2006)
6. Agarwal, S., Katz, R., Krishnamurthy, S.V., Dao, S.K.: Distributed Power Control in Ad Hoc Wireless Networks. Proceedings of the IEEE international symposium personal, indoor, and mobile radio communication (PIMRC), pp. F59–F66, 2001
7. Nandi, A., Kundu, S.: Energy level performance of retransmission schemes in wireless sensor networks over rayleigh fading channel. Proceedings of the IEEE international conference on computational intelligence and communication networks (CICN 2010), pp. 220–225, 2010
8. Nandi, A., Kundu, S.: Evaluation of optimal transmit power in wireless sensor networks in presence of rayleigh fading. ICTACT J Commun Technol 1(2), 107–112 (2010)
9. Cruz, R.L., Santhanam, A.V.: Optimal routing, link scheduling and power control in multi-hop wireless networks. Proc. IEEE Conf. Computer Comm. 1, 702–711 (2003)
10. Nandi, A., Kundu, S.: Energy level performance of packet delivery schemes in wireless sensor networks in shadowed channel. Sens. Transducers J. 118(7), 73–86 (2010)
11. Narayanaswamy, S., Kawadia, V., Sreenivas, R.S., Kumar, P.R.: Power control in ad hoc networks: theory, architecture, algorithm and implementation of the COMPOW protocol. Proceedings of the European wireless next generation wireless networks: technologies, protocols, services, and applications, pp. 156–162, 2002
12. Stanojev, I., Simeone, O., Bar-Ness, Y., Kim, D.H.: IEEE Trans. Wireless Commun. 8(1), 326–335 (2009)
13. Rossi, M., Zanca, G., Stabellini, L., Crepaldi, R., Harris, A.F., Zorzi, M.: Synapse: A network reprogramming protocol for wireless sensor networks using fountain codes, 5th annual IEEE communications society conference on sensor, mesh and ad hoc communications and networks, pp. 188–196, 2008
14. Sesia, S., Caire, G., Vivier, G.: Incremental redundancy hybrid arq schemes based on low-density parity-check codes. IEEE Trans. Commun. 52(8), 1311–1321 (2004)
15. Nosratinia, A., Hunter, T., Hedayat, A.: Cooperative communication in wireless networks. IEEE Commun. Mag. 42(10), 68–73 (2004)
16. Hunter, T., Nosratinia, A.: Diversity through coded cooperation. IEEE Trans. Commun. 5, 283–289 (2006)
17. Nandi, A., Kundu, S.: Optimal transmit power and packet size in wireless sensor networks in shadowed channel. Int. J. Sens. Networks 11(2), 81–89 (2012)
18. Ferrari, G., Tonguz, O.K.: Performance of ad hoc wireless networks with aloha and PR-CSMA MAC protocols, Proceedings of the IEEE Global Telecommunication Conference (GLOBECOM), pp. 2824–2829, Dec 2003
19. Goldsmith, A.: Wireless communications. Cambridge University Press, Cambridge (2005)
20. Sklar, B.: Rayleigh fading channels in mobile digital communication systems Part I: characterization. IEEE Communication Magazine, pp. 90–100, July 2003
21. Shih E., Cho, S.-H., Ickes, N., Min, R., Sinha, A., Wang, A., Chandrakasan, A.: Physical layer driven protocol and algorithm design for energy-efficient wireless sensor networks. In Proceedings of the 7th annual international conference on mobile computing and networking (MobiCom '01), 2001
22. Intel® StrongARM® SA-1110 Microprocessor Developer's Manual, Intel Corporation, http://www.intel.com, 1999
23. Forney, G.D.: The Viterbi algorithm. Proc. IEEE 61(3), 268–278 (1973)
24. Heller, J.A., Jacobs, I.M.: Viterbi decoding for satellite and space communication. IEEE Trans. Commun. Technol. 19(5), 835–848 (1971)
25. Yasuda, Y., Kashiki, K., Hirata, Y.: High rate punctured convolutional codes for soft decision Viterbi decoding. IEEE Trans. Commun. 32(3), 315–319 (1984)

26. Haccoun, D., Begin, G.: High-rate punctured convolutional codes for Viterbi and sequential decoding. IEEE Trans. Commun. **37**(11), 1113–1125 (1989)
27. Kleinschmidt, J.H., Borelli, W.C., Pellenz, M.E, An analytical model for energy efficiency of error control schemes in sensor networks. ICC '07. IEEE International Conference on Communications 2007, pp. 3895–3900, 2007
28. Nandi, A., Kundu, S.: Energy efficient packet data service in wireless sensor network in presence of raylrigh fading. Int. J. Grid High Perform. Comput. (IJGHPC) **3**(3), 31–44 (2011). doi:10.4018/jghpc.2011070103

Trust-Based Routing for Vehicular Ad Hoc Network

Suparna DasGupta and Rituparna Chaki

Abstract Vehicular ad hoc networks are likely to become the most relevant form of mobile ad hoc networks with special requirements in terms of node mobility and comprise of vehicle-to-vehicle (V2V) and vehicle-to-infrastructure (V2I) communications. The deployment of vehicular communication systems is strongly dependent upon their underlying security and privacy features. The effective trust management schemes for VANETs have been given the dire consequences of acting on false information management. The urgent nature of communication necessitates that messages should be signed and verified before they are trusted and it should be done to keep secrecy of vehicles real identity. Prerequisite to communicate within VANETs is an efficient route between network nodes which must be adaptive to the rapidly changing topology of VANET. In this paper, we have proposed a new trust-based efficient routing protocol for VANETs and provide the solution for avoiding the channel congestion and performance bottleneck problem. Conducted simulation experiments on different scenarios show the performance analysis and effectiveness of the new proposed routing protocol for vehicular ad hoc networks.

Keywords Vehicular ad hoc networks · Security · Privacy · Trust · Congestion

S. DasGupta (✉)
Department of Information Technology, JIS College of Engineering, Kalyani, West Bengal, India
e-mail: suparnadasguptait@gmail.com

R. Chaki
A.K. Choudhury School of Information Technology, University of Calcutta, Kolkata West Bengal, India
e-mail: rituchaki@gmail.com

© Springer India 2015
R. Chaki et al. (eds.), *Applied Computation and Security Systems*, Advances in Intelligent Systems and Computing 304, DOI 10.1007/978-81-322-1985-9_13

1 Introduction

Vehicular ad hoc networks is a wireless network that is formed between vehicles on demand basis and have become a popular area for both the academic research community and automobile industry, with specific attention to improving driving experience and road safety. As the vehicles change their location constantly, there is a continuous demand for information on the current location and specifically for data on the surrounding traffic, routes, and much more.

In vehicle-to-vehicle (V2V) communication, three broad categories of architecture are related, such as infrastructure-based, ad hoc networks, and hybrid. The infrastructure-based architecture takes advantage of the existing cellular networks. This network has few drawbacks as: high operation cost, limited bandwidth, and symmetry channel allocation for uplink and downlink. As infrastructure do not required in ad hoc networks, the cost of building such network will be very low and it can even operate in the events of disasters. The hybrid architecture combines these two architectures by considering vehicles as data relays between roadside base stations. This architecture also requires the function of multi-hop communication between vehicles, which is the essential part of ad hoc network architecture.

VANET consists of vehicles and road-side units as network nodes and enables inter-vehicle communication or IVC along with the road side-to-vehicle communication, i.e., RVC. Road conditions such as congestion, collisions, or constructions are shared by vehicles through VANETs. IVC and RVC can be divided into two categories, such as: safety-related applications and infotainment applications. Besides the fundamental security requirements, sensitive information, i.e., identity and location privacy should be preserved; on the contrary, traceability is required where the identity information needs to be revealed. In addition, privilege revocation is required by network authorities. V2V and vehicle-to-infrastructure (V2I) communication can enable a range of applications to enhance transportation safety and efficiency as well as infotainment.

VANETs face many interesting research challenges in multiple areas, from privacy and anonymity to the detection and eviction of misbehaving nodes. Securing vehicular communication is a tough problem due to tight coupling between application and the network fabric, as well as additional social, legal, and economical consideration, which raise a unique combination of operational and security requirements. Privacy and security are important issues in vehicular networks. Users wish to maintain location privacy and anonymity, location and direction of movement of the vehicle are known only to those legally authorized to have access to them and remain unknown to anybody unauthorized. Security-related key challenges for VANETs are control access, user authentication, message authentication, message integrity, message identification, message privacy, accountability anonymous certification, group signatures, PKI: managing certificate revocation, pseudonyms, etc. In case of designing any routing algorithm for VANET, the above-discussed issues should be taken into account.

In this paper, we have introduced a new reliable communication mechanism depending on a proposed system module for VANET. This proposed scheme consists of two different steps. (1) Registration procedure has been introduced for new vehicles, and trust value has been assigned to each of the registered vehicles. (2) Communication mechanism has been presented for existing vehicles.

The rest of the paper is organized as follows. Comprehensive surveys of related works of different secure routing protocols for VANETs are discussed in Sect. 2. In Sect. 3, we have presented new trust-based routing for VANET. Intensive performance analysis of our proposed scheme is presented in Sect. 4. We conclude our paper with final remarks in Sect. 5.

2 Related Works

For full deployment of VANETs, two paramount issues should be resolved, namely security and privacy. The information communicated by vehicles should be secured. Many researchers have been already published number of research papers, addressing the security issue of vehicular ad hoc networks. In this section, we have discussed some of the security-related research challenges of VANET.

In VANETs, the connection between two vehicles is often intermittent due to dynamic vehicular movement. Routing in VANETs has been studied, and many different protocols were proposed. Granelli et al. proposed a motion-based routing algorithm [1] for VANET. The routing metric enables to exploit not only positioning information but also the direction the vehicles movement. Extensive evaluation outlines the advantages of MORA [1], especially in case of high mobility of vehicles and frequent topology changes. But considering these parameters are not enough for a best next hop selection in VANETs. A vehicle that is almost out the communication range should not be selected as a next hop, which cannot be guaranteed without taking into account the speed. Menouar et al. [2] proposed MOPR, taking into account neighboring vehicles movement speed additional to MORA [1]. Vehicle that is estimated to go out the communication range in a short-duration time will not be selected as a next hop. This approach helps in minimizing the risk of broken links and in reducing data loss. The performance of the scheme largely depends on the prediction accuracy and the estimate of the transmission time that depends, in turn, on several factors such as network congestion status, driver's behavior, and the used transmission protocols. In Kumar and Rao [3] proposed a position-based greedy routing protocol, which uses the location, speed, and direction of motion of their neighbors to select the most appropriate next forwarding node. Like GPSR [4], it uses the two forwarding strategies greedy and perimeter. It predicts the position of nodes within the beacon interval whenever it needs to forward a data packet. DGRP [3] selects better next hop node than MOPR [2] as MOPR selects only those nodes for forwarding which are going to be communication range for next one second. In [3], if link stability between the forwarding node and its neighbor node is weak, possibility of packet loss is high in DGRP [3] and also prediction of position information is not reliable at all instances. Gong et al. [5] proposed a predictive directional greedy routing protocol, in which the weighted score is calculated from two strategies namely, position first forwarding and direction first forwarding. Using these strategies, the current neighbors and possible future neighbors of packet carrier are found. In PDGRP [5], next hop selection is done based on prediction and it is not reliable at all situations. In Jayasudha and Chandrasekhar [6] proposed a hierarchical cluster based greedy routing protocol. The main objective of the algorithm is to optimize the packet behavior in ad hoc networks with high mobility and to deliver messages with high reliability. We have also proposed a routing solution in [7]. In [8], J. Serna et al. proposed a geo-location-based trust for VANET's privacy and used as an authorization paradigm based on a mandatory access model and a novel scheme which propagates trust information based on a vehicle's geo-location. A trust-based privacy preserving model for VANETs has been presented by Ayman Tajeddine et al. [9], which is unique in its ability to protect privacy while maintaining accurate reputation-based trust. In [10], a reputation based trust model has been presented by Qing Ding et al. This is an event based reputation model to filter bogus warning messages. A dynamic role dependent reputation evaluation mechanism has been presented to determine whether an incoming traffic message is significant and trustworthy to the driver. BROADCOMM [11] is a popular broadcast algorithm for emergency situation in VANET. By exchanging Hello message to neighbor nodes, all nodes determine their own cell boundary. A cell reflector node is selected for each cell. These nodes actually relay the broadcasted message from one cell to other cell. Its' simple to implement nature is ideal in emergency alert system.

The above discussions lead to the conclusion that is mainly cryptographic, and certificate-based techniques [12] are being preferred by the researchers for securing communication within VANETs. This leads to another problem namely, certificate revocation problem [13].Some researchers have chosen trust value-based authentication, but the parameters influential in trust value assignment of a vehicle are not properly identified. A trust based solution in this issue is also proposed by us [14]. Maintaining pseudonym [15] is another approach for achieve security. But it also leads to an extra maintenance cost. This paper aims to provide a reliable communication mechanism for vehicular ad hoc network, and the proposed solution is used to overcome performance bottleneck and channel congestion problem.

3 Trust-Based Routing Protocol for Vehicular Ad Hoc Networks

In this section, we are going to propose a solution of above-discussed problem. In our proposed solution, we have distributed VANET in a layered architecture. In the lowest layer, all nodes (i.e., vehicles) present in the system. Local registration authority (LRA_i) implies a road-side unit that acts as a middle layer element within

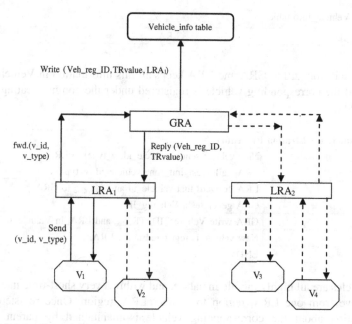

Fig. 1 Modular diagram of the system

the framework. LRA$_i$ is responsible for maintaining vehicles registered under it. The highest layer component, called global registration authority (GRA), is nothing but a repository having all lower layer information (Fig. 1).

3.1 Registration Procedure

Here, we have assumed that all nodes in a VANET are distributed according the proposed layered architecture. On entry of a new vehicle in the system, it sends a request for a registration certificate to its LRA$_i$, i.e., LRA$_i$ in its range and this request is termed as registered to communicate (RTC). LRA$_i$ estimates trust value of that vehicle forwards it and vehicle number to GRA. GRA generates a unique sequence number, i.e., USN for that particular vehicle. This USN acts as Veh_reg_id for the corresponding vehicle. At the registration time, LRA$_i$ do not know the behavior of the vehicle. For this reason at this time, an initial trust value is given to the vehicle.

- **Trust Value Initialization**

In this subsection, we have been presented an algorithm for new vehicle entering in the system. Every new vehicle has to register under its local LRA$_i$. For this reason, it sends a registration request to LRA$_i$. In this request, each vehicle has to send their types and the unique features of it. After receiving the request, LRA$_i$ assigns a unique number and a trust value to the requesting vehicle. After initialization of trust

Table 1 Vehicle_info table

Veh_reg_ID	LRA$_i$	TR$_{value}$

value, it will forward to GRA and GRA keeps all this information in Vehicle_info table and the corresponding vehicle is registered under the communicating LRA$_i$ (Table 1).

Algorithm 1: Registration Procedure	
Step 1:	New vehicle sends (vehicle_id, v_type) to LRA$_i$
Step 2:	LRA$_i$ call Trust_init_func(vehicle_id, v_type)
Step 3:	LRA$_i$ forward that vehicle_id and TR$_{value}$ to GRA
Step 4:	GRA generates a Veh_reg_ID
Step 5:	GRA write Veh_reg_ID, TR$_{value}$ and LRA$_i$ in Vehicle_info table.
Step 6:	New vehicle is registered under LRA$_i$
Step 7:	END.

Vehicles are all highly mobile in nature, and within a very short-time interval, it can move from one LRA$_i$ region to another LRA$_i$ region. Once registered, all information about the corresponding vehicle is maintained by parent LRA$_i$. Information of every registered vehicle's is also stored in GRA, and GRA actually acts as a global repository of all registered vehicles. LRA$_i$ monitors all vehicles registered under it. When a vehicle moves out of its region, it broadcast a message consisting information about that vehicle. In this way, the new LRA$_i$ in which region the vehicle enters can know information about it. The information sends by the following message format (Fig. 2).

In the above-message format, FM$_i$ denotes an identifier that uniquely identifies the message. Veh_reg_id, TR$_{value}$, and Parent_LRA$_i$ denote registration identifier, trust value, and initial LRA$_i$, respectively, for the corresponding vehicle. In this way, when LRA$_i$ found any new vehicle in its region, it also had some essential information about that vehicle. If LRA$_i$ needs more detail information, then it can query to GRA and gets required information from it.

At the execution of all the above-mentioned steps, two kinds of problem can be occurred as; performance bottleneck problem for LRA$_i$ and channel congestion problem.

- **Performance Bottleneck Problem**

LRA$_i$ is busy for registering new vehicles, estimating there trust value in a certain time interval, sending frequent message to other LRA$_i$, etc. All these jobs have done for a large number of vehicles depending locality. As a result large

FM$_i$	Veh_reg_ID	TR$_{value}$	Parent_LRA

Fig. 2 Message format

number of jobs may be waiting in a queue. For this reason, priority should be assigned to each job for job scheduling. In this way, high-priority job performed first and low-priority job have to wait. In this way, a starvation problem has been occurred for low-priority job. To solve this problem, aging technique can be introduced.

- **Channel Congestion Problem**

For performing assign jobs, LRA_i has to communicate with each other and with vehicles. Vehicles also communicate with each other for data transmission. All these communication take place using message passing through network channels. Due to this large number of data transmission, channel congestion can occur which cause to data loss. For avoiding this problem, channel availability should check at link layer. When a node wants to communicate, it sends a request to send (RTS) and waits until it received an ACK message. We can assume for this time this node become busy. On the other hand, when a node receives RTS, it also become busy until it sends ACK message. For other nodes, when RTS or clear-to-send (CTS) is received (but it is not send by themselves), they can assume that for that time period channel will be used. And, we assume this time period is a specified time period, as network transmission time (NTT). For calculation of this NTT, we can take one of the two following techniques.

- **No Persistent Technique**

In this technique, messages send after a certain time interval. This interval value can be estimated depending on some network property. By a thorough survey, we have assessed that this time interval depends on maximum transfer unit (MTU) of a network. A relation between channel capacity and NTT is also found. From our observation, we can write

$$MTU * CC \propto 1/NTT$$
$$MTU * CC = K /NTT, \quad \text{where K is a constant} \qquad (1)$$
$$NTT = K/MTU * CC$$

That means other nodes have to wait for this specified time period before starting the requesting process to access channel.

- **N-persistent Technique**

In this technique, messages send depending on a probability of $n\ \%$ success. The probability of failure is $(1 - n)\ \%$.

3.2 Communication Procedure

- Assumptions
- All LRA_i' maintain information about their child nodes and store in a list, defined child_list {Veh_reg_ID, address, TR_{value}}.

- LRA$_i$ maintains information about vehicles in its one-hop distance.
- All vehicles registered under LRA$_i$, are in a one-hop distance of each other.

Communication taken place in this type of network can be categorized into two types.

- **Query Driven**

This type of communication is reactive in nature. A node broadcasts the query to procure necessary information. Then, it waits for T_{qb} time period, where

$$T_{qb} = \text{pkt}_{size} * \text{NTT} * 2 * T_{range} \tag{2}$$

The communication completes successfully if reply from any node is received within the time. If no such information is received, the LRA$_i$ of sender vehicle multicast its query to all LRA$_i$ and starts the timer for T_{qm} time period, where,

$$T_{qm} = \text{pkt}_{size} * \text{NTT} * 2 * \text{LRA}_{dist} \tag{3}$$

After this specific time period, sender LRA$_i$ checks for reply. If any reply found, then it forwards to the sender vehicle node. Otherwise communication declared as a failure.

Algorithm 2: Query-driven Communication	
Let V1 requires some information. So, it starts a communication session.	
Step 1:	V1 broadcasts a query and waits for time T_{qb} periods, where $T_{qb} = \text{pkt}_{size} * \text{NTT} * 2 * T_{range}$
Step 2:	If reply comes, then go to step 6
	Else go to next step.
Step 3:	V1 requests it's LRA$_{v1}$ to forward it's query.
Step 4:	LRA$_{v1}$ multicasts this query to other LRAs and waits for time T_{qm} periods, where $T_{qm} = \text{pkt}_{size} * \text{NTT} * 2 * \text{LRA}_{dist}$
Step 5:	If reply comes, then forward to V1
	else sends a failure message to V1
Step 6:	Communication successful.

- **Specific Vehicle-to-Vehicle Communication**

This type of communication takes place when any vehicle wants to communicate with any other specific vehicle. A prerequisite of this type of communication is a logical route establishment between sender and receiver vehicle. In the beginning of the communication, sender vehicle initiates a query about address of receiver vehicle. Sender vehicle search for receiver vehicle, in LRA$_i$ child_list{Veh_reg_ID, TR$_{value}$}. If found then directly communicate with that specific vehicle. Otherwise, sender vehicle's parent LRA$_i$ send message to all LRA$_i$. The receiver vehicle should be enlisted in the child list of any of the LRA$_i$.

Table 2 Data Dictionary

Parameter	Details
LRA	Local registration authority
GRA	Global registration authority
RTC	Registered to communicate
USN	Unique security number
Veh_reg_ID	Vehicle registration identifier
TR_{value}	Trust value
Vehicle_id	Vehicle identifier
V_type	Vehicle type
FM_i	Frequent message identifier
T_o	Time of a vehicle entering in system
T_{mr}	Total number of request send within time period
T_{mn}	Total number of reply received within time period
MTU	Maximum transfer unit

Unavailability of reply is suggestive of either the specific node has roamed to a different node or it is temporarily down. After receiving the reply, sender node can start communicating with receiver node (Table 2).

Algorithm 3: Routing Procedure	
Let S is the sender vehicle and D is the receiver one.	
Step 1:	Sender vehicle S searches in it's LRA_i child list for destination vehicle.
Step 2:	If destination vehicle found then, go to step 5
	Otherwise send communication request to it's LRA_i
Step 3:	LRA_i sends a request message to other LRA_i for searching destination address.
Step 4:	If found, then send it back to source node, Otherwise send a failure message to sender vehicle.
Step 5:	Send a success message to sender vehicle.

4 Performance Analysis

In order to implement the above-discussed proposal, the system performance needs to be tested in real life, despite the many obstacles that make this difficult such as the expense, high mobility, network complexity, and distributed environments. Simulation tools are considered the best means with which to evaluate the performance of any network type particularly wireless and ad hoc networks. For example, this method enables the user to emulate the network in terms of routing protocols, security constraints, and other factors that are similar to real-life situations, thus avoiding the difficulties resulting from the existence of obstacles.

Table 3 Simulation environment parameters

Parameter	Value
Channel type	Wireless channel
Radio-propagation model	Two-ray ground
Antenna model	Omni antenna
Network interface type	Wireless Phy
Mac type	802.11
Number of nodes	25

We choose the NS2 simulator for this analysis because it realistically models arbitrary node mobility as well as physical radio-propagation effects such as signal strength, interference, capture effect, and wireless propagation delay. Our propagation model is based on the two-ray ground reflection model. The simulator also includes an accurate model of the IEEE 802.11 Distributed Coordination Function Wireless MAC protocol. Using NS2, we evaluate the performance of the proposed protocol and present the following metrics for comparing the performance with the different well-known traditional routing. The simulation model consists of a network model that has a number of wireless nodes, which represents the entire network to be simulated (Table 3).

We have to examine whether the proposed routing algorithm works robustly or not. For this reason, we have compared our routing algorithm with LAR, DSR, AODV, and GPSR according to the following metrics.

(i) Packet delivery ratio: Measures the ratio of data packets delivered to the destinations and the data packets generated by the CBR source.

Say, N is the number of packet generated by the CBR source. Among those D packets are received by destination node

$$\text{Packet delivery ratio, PDR} = D/N \qquad (4)$$

This number indicates the effectiveness of a protocol.

(ii) End-to-end delay: Measured in milliseconds, includes processing (pd), route discover latency (rl), queuing delays (qd), retransmission delay (rd) at the MAC, and propagation (pr) and transmission times (tr). This number measures the total delay time from a sender to a destination. So, we can compute

$$\text{End-to-end delay} = \text{pd} + \text{rl} + \text{qd} + \text{rd} + \text{pr} + \text{tr} \qquad (5)$$

(iii) Normalized routing load (NRL): Measures the number of routing packets transmitted per distinct data packet delivered to a destination. Let DP is the number of data packet and for delivering these; we have required CP number of routing packet.

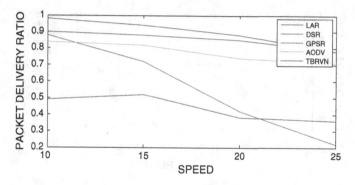

Fig. 3 Packet delivery ratio versus speed

$$NRL \ = \ CP/DP \qquad (6)$$

The routing overhead is an important metric for comparing these protocols as it measures the scalability of a protocol, and its efficiency in terms of throughput and power consumption. In our simulation study, we have performed sensitivity analysis to investigate the effect of various network parameters.

- **Effect of Speed**

This study is based on 100 nodes with 10 communication sessions. We have set our simulation with zero pause time to stress the mobility in the network. To understand the effect of speed on performance, we varied the speed of the vehicles between 10 m/s (or 22 miles/h) and 25 m/s (or 56 miles/h). The simulation results are presented in Figs. 3, 4 and 5. They show performance trade-off in some techniques.

Though AODV, LAR, and our proposed algorithm deliver almost similar number of packets irrespective of mobility, AODV and LAR have high end-to-end delay and control packet overhead. In a highly mobile scenario, links tend to break frequently. In such situation, these two algorithms need to send more route discovery message. LAR also suffers from inaccurate prediction of the request zone, which leads to network-flooding problem. DSR performs similar to our proposed protocol in terms of end-to-end delay and number of control packet transmitted per data packets. However, it gives poor result with respect to packet delivery ratio. This is because DSR has to rediscover routes more frequently as vehicle speed increases. In case of GPSR, high-control overhead is caused by maintaining neighbor location, and high end-to-end is caused by the outdated neighbor information. This causes GPSR to forward to non-existing neighboring nodes. From these results, we can conclude that connection-oriented approaches either drop a large amount of data packets (for example, DSR) or require a large number of control packet to keep routes up-to-date (for example, AODV and LAR) and

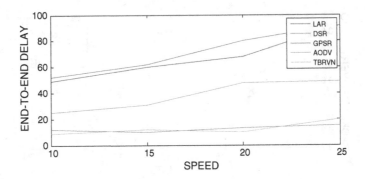

Fig. 4 End-to-end delay versus speed

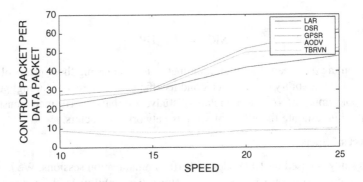

Fig. 5 Control packet per data packet versus speed

neighboring nodes information (for example, GPSR). In our proposed one, all vehicles' information is stored in vehicle_info table. This is maintained by a centralized authority, GRA. From this any vehicle's location is accessible. In addition, when a vehicle leave a region of an LRA_i (because of its' speed) and enters another LRA's region, then through frequent messaging 2nd LRA_i becomes aware about this particular vehicle. For this reason, despite of speed increase proposed protocol gives a consistence result.

- **Effect of Network Density**

In this case, we assumed node mobility is 25 m/s. For testing the effects of network density, we simulated with 50, 100, 150, 200, and 400 nodes. The results of analysis are plotted in Figs. 6, 7 and 8. In terms of packet delivery, AODV and LAR perform initially better than our proposed protocol. From our simulation result, it is clearly visible that with the increase of network density our protocol starts to perform better than AODV and LAR. If we compare on the basis of end-to-end delay obviously our proposed protocol perform best than others. Due to increment of number of nodes, number of control packets also increases for AODV

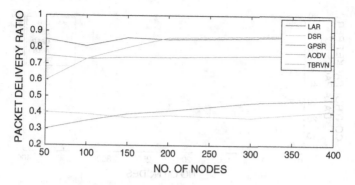

Fig. 6 Packet delivery ratio versus number of nodes

and LAR. DSR is not performed well. For GPSR, increasing number of nodes means more number nodes to maintain. So, route selection procedure becomes more time-consuming. This situation is reflected in the following figure. Though there is a centralized repository maintained by a centralized authority, the total responsibility of routing is actually distributed among LRA's. The LRA's are distributed locationwise. Thus, increases in the number of nodes are not going to reflect very much in case of proposed protocol.

From the above different comparisons we can see that the proposed routing protocol provides the better result than traditional well-known routing.

We have also examined our protocol's functionality with respect to BROAD-COMM [11], a routing protocol specially aimed at vehicular communication. BROADCOMM [11] is chosen due to its' easy-to-implement features which have made it a popular choice for routing in VANET. The metrics for comparison between our proposed technique and BROADCOMM are as follows: packet delivery ratio, end-to-end delay, and routing load (Fig. 9).

It has been observed that performance of BROADCOMM [11] is better than the proposed technique. The reason is BROADCOMM does not involve any checks on the trust worthiness of the vehicles involved in communication. Our algorithm has

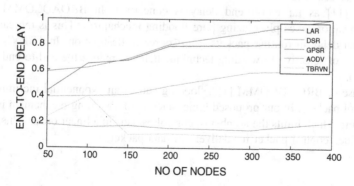

Fig. 7 End-to-End delay versus number of nodes

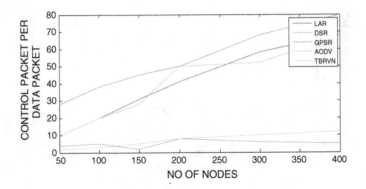

Fig. 8 Control packet per data packet versus number of nodes

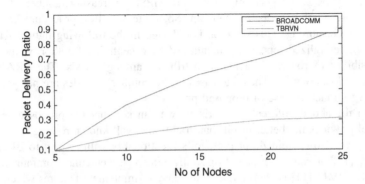

Fig. 9 Packet delivery ratio versus number of nodes

to perform certain mandatory checks and that adds to the delay in packet delivery ratio. Thus, in an ideal situation, BROADCOMM [11] proves to be a better performer (Fig. 10).

The performance of our proposed logic is much better than that of BROAD-COMM [11] as far end-to-end delay is concerned. In BROADCOMM [11], communication takes place using pure flooding mechanism. This is the cause of additional delay in routing packets from source to destination. In TBRVN algorithm, use of selective forwarding technique helps to reduce the end-to-end delay (Fig. 11).

In case of BROADCOMM [11], flooding causes an exponential incrimination of control packets. In our proposed logic, selective forwarding mechanism is used for routing which limits the number of control packets to a linear order. This result is to reduce control packet for delivery of data packet.

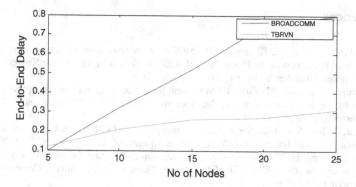

Fig. 10 End-to-end delay versus number of nodes

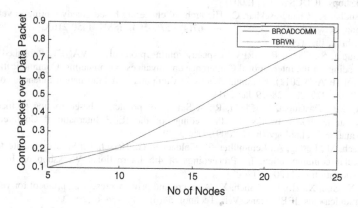

Fig. 11 Control packet per data packet versus number of nodes

5 Conclusions

In this paper, we have presented a reliable routing protocol for vehicular ad hoc network. Here, we have proposed a trust-based registration mechanism to provide the reliability. A layered structure has been presented for the authenticate vehicles communication. Performance bottleneck and channel congestion problem also have taken care consideration in this proposal. The results show that proposed routing protocol provides better result compared with the existing well-known routing protocol. We have also compared our proposed routing algorithm with BROADCOMM, a routing algorithm specially aimed at VANET environment. Results proved that TBRVN has less delay and routing overhead as compared with existing routing protocols.

References

1. Granelli, F., Boato, G., Kliazovich, D.: MORA: A movement-based routing algorithm for vehicle ad hoc networks. In: Proceedings of IEEE Workshop on Automotive Networking and Applications (AutoNet 2006), San Francisco, Dec 2006
2. Menouar, H., Lenardi, M., Filali, F.: Movement prediction based routing concept for position based routing in vehicular networks. In: Proceedings of the 66th IEEE Vehicular Technology Conference, 30 Sept–3 Oct 2007
3. Kumar, R., Rao, S.V.: Directional greedy routing protocol (DGRP) in mobile ad hoc network. In: Proceedings of International Conference on Information Technology, ICIT (2008)
4. Karp, B., Kung, H.T.: GPSR: greedy perimeter stateless routing for wireless networks. In: Proceedings of the ACM/IEEE International Conference on Mobile Computing and Networking (MobiCom) (2000)
5. Gong, J., Xu, C., Holle, J.: Predictive directional greedy routing in vehicular ad hoc networks. In: Proceedings of 27th International Conference on Distributed Computing Systems Workshops (ICDCSW '07) (2007)
6. Jayasudha, K., Chandrasekhar, C.: Hierarchical clustering based greedy routing in vehicular ad hoc networks. Eur. J. Sci. Res. **67**(4), 580–594. ISSN 1450-216X, Euro Journals Publishing, Inc. (2012)
7. DasGupta, S., Chaki, R.: SRPV: a speedy routing protocol for VANET. Published in the Proceedings of the International Conference on Advances in Computing, Communication and Control (ICAC3-2011), Communications in Computer and Information Science, vol. 125, Part 2, pp. 275–284, 28–29 Jan 2011
8. Saha, S.B., DasGupta, S., Chaki, R.: A Survey of prediction based routing protocols for vehicular ad hoc networks. In: Proceedings of the IEEE International Conference on Information Technology (ICIT-2009), Dec 2009
9. Gerlach, M., Festag, A., Leinmüller, T., Goldacker, G., Harsch, C.: Security architecture for vehicular communication. In: Proceedings of 4th International Workshop on Intelligent Transportation (WIT2007) (2007)
10. Lin, X., Sun, X., Ho, P.H., Shen, X.: A secure and privacy preserving protocol for vehicular communications. IEEE Trans. Veh. Technol. **56**(6), 3442–3456 (2007)
11. Durresi, M., Durresi, A., Barolli, L.: Emergency broadcast protocol for inter-vehicle communications. In: Proceedings of 11th IEEE International Conference on Parallel and Distributed Systems (ICPADS'05)
12. Papadimitratos, P., Buttyan, L., Hubaux, J.P., Karg, F., Kung, A., Raya, M.: Architecture for secure and private vehicular communications. In: Proceedings of 7th International Conference on ITS Telecommunications (ITST'07), pp. 1–6, June 2007
13. Raya, M., Jungels, D., Papadimitratos, P., Aad, I., Hubaux, J.P.: Certificate revocation invehicularNetworks. http://citeseerx.ist.psu.edu/viewdoc/summary?doi=10.1.1.92.2291S. 2006
14. DasGupta, S., Chaki, R., Choudhury, S.: TruVAL: trusted vehicle authentication logic for VANET. Published in the Proceedings of the International Conference on 3rd International Conference on Advances in Computing, Communication and Control (ICAC3-2013) in Mumbai, 18th–19th Jan 2013
15. Wiedersheim, B., Ma, Z., Kargl, F., Papadimitratos, P.: Privacy in inter-vehicular networks: why simple pseudonym change is not enough. In: Proceedings of 7th International Conference on Wireless On-demand Network Systems and Services (WONS), pp. 176–83 (2010)

Study on Handover Mechanism in Cellular Network: An Experimental Approach

R.K. Mishra, Nabendu Chaki and Sankhayan Choudhury

Abstract Mobility of devices in cellular network induces several problems. This has drawn reasonable research attention in recent times. Designing appropriate Call Admission Control (CAC) mechanism and evaluating performances for such algorithms are among the crucial challenges in the domain. Such technologies continually evolve with the changing communication paradigms. In this paper, we have investigated various parameters that impact performance at home base station to facilitate seamless handover by an exhaustive simulation using George Mason University's Java-based MASON framework. The different parameters considered during simulation include handover threshold limits, user velocity, process prioritization, number of users in a cell, and redundancy for transceiver for better coverage. Through this exhaustive simulation, this paper aims to estimate the effect of factors influencing handoff as well as to test existing hypothesis about the handoff. The experimental results have disproved certain general conventions that are often assumed.

Keywords Handover · RACH · Cell identification broadcast · Microcell · Macrocell · Beaconing · Call drop

R.K. Mishra (✉)
Feroze Gandhi Institute of Engineering and Technology, Raebareli, India
e-mail: rakesh.mishra.rbl@gmail.com

N. Chaki · S. Choudhury
Department of Computer Science and Engineering, University of Calcutta, Kolkata, India
e-mail: nabendu@ieee.org

S. Choudhury
e-mail: sankhyan@gmail.com

© Springer India 2015
R. Chaki et al. (eds.), *Applied Computation and Security Systems*, Advances in Intelligent Systems and Computing 304, DOI 10.1007/978-81-322-1985-9_14

1 Introduction

Wireless Cellular network has expanded its outreach in every aspect from its accessibility to various functionalities. The network started as alternative to provide connectivity at remote and hostile areas where wired network was impossible to expand. The proven utility of the network in such area has made it an alternative mode of communication for people on the move. The cellular network grows from alternative mode of communication to main stream backbone network. It has no more remained a mode of nomadic communication; rather, it transformed into necessity of modern civilization. Wireless communication is not only used for vocal communication but also for data accessing. The data usage through cellular networks has grown by several folds in past few years. The growing demand is continuously thrusting for more resource provisioning and finer-tuning of components in network.

Over time, wireless networks have evolved into five generations. Interestingly, the newer generations did not replace the older and these are coexisting. Wireless networks, irrespective of its generation and technology, like any other technological innovation also posses certain challenges. These challenges may be inherent to the network or arising out of its massive deployment. The inherent challenges with the wireless network are handoff commissioning, channel encoding, bandwidth utilization, interference management, data security, etc. Other technological challenges such as interconnectivity, frequency reuse, coding, traffic management, resource scheduling, channel multiplexing are due to the massive deployment.

All the above challenges are continuously being studied, and solution from different perspectives is proposed. However, the area is still quite open for further exploration. Mobility is an essential characteristic of wireless communication. In order to provide seamless connectivity over a larger geographical area, the entire space is divided into smaller coverage area called cells. A cell can also be referred as a service area of the network served by single transceiver of the network. The handover is a process of transferring of a mobile user from current service area to adjacent service area and maintain a seamless connectivity [1, 2].

Success of any wireless network or service provider is often determined by how efficiently handover is managed and executed. The problem of handoff is as old as wireless networks, but still as new as next upcoming standard in arena. This paper aims to estimate the effect of factors influencing handoff as well as to test existing hypothesis about the handoff using simulation experiments. It is also studied to understand how these parameters may be used in the perspective of microcell mobility in urban mobility patterns and off-loading the base station from routing chores by increasing participation of mobile nodes.

2 State-of-the-Art Review

Handover enables any mobile communication system to have a seamless avail-ability of service during mobility. In order to support a handover process, band-width is reserved for migration of a mobile unit from service area to another. The bandwidth reservation for handover can be either a static or a dynamic. In static partitioning, predefined number of channels is designated for the handover and their count remains unaltered. Main pitfall of static partitioning scheme is that it is not adaptable to voluminous handovers. In fact, in such a situation, the requests will be denied and call will be dropped. On the contrary, dynamic channel allo-cation scheme is adaptable to varying demand of handover as it can hire channel(s) through any means for handover. It suffers from the shortcoming like gradual acquisition of maximum bandwidth thus deprivation of other users for resource allocation [1]. None of the solutions or solution variants attempt to study the affecting parameters for the handover process [3].

In [4], the authors have mentioned the relevance of various parameters, such as received signal level, received signal quality, distance, and the corresponding thresholds for sustaining the communication. Experiments in [4] show that men-tioned parameters play crucial role in the handover decision making.

In [5], a comprehensive review of state of the art for handover techniques hovers around mobile unit mobility and other associated parameters. The review exhaustively covers the handover on the basis of usage, prioritization, channel reservation, and mobility. It is a compilation of Call Admission Control (CAC) techniques on the predicted movement of mobile unit using both RSS value and past history.

Group handover strategies (with and without access point) has been discussed in [6] and a proposal for group handover without using access point. The authors have proposed that all the devices collect together within a small network by initiating any of the standard neighbor discovery protocol. Thus, knowing the count and resource requirement appropriate access point can be identified and request for handover is initiated. The authors have computed the expression for node count and other metrics using optimization equations. Thus, the overhead for unwarranted computation increases at main switching center toward deploying such solution.

Irrespective of the technology wireless communication system divides geo-graphical area into smaller units called cell. Each cell is categorized by its dedi-cated serving transceiver called base station. Over the period of existence within a cell, mobile stations (MSs) continuously measure the signal strength of adjoining base station including home base station and send log to home base station. Base station maintains a log of periodically measured signal strength from each MS. As the signal strength of any of the adjoining base station reaches a handover threshold for a MS, a request for handover is initiated by serving base station for new base station. Request is forwarded up in the network hierarchy, if required. In case of denial, a call is queued/dropped; otherwise, host base station is acknowledged to start handover. Handover is executed between the two predefined limits of the signal strength called as thresholds, and to avoid ping-pong effect, a reference value called hysteresis is used. After commencement, all the databases and network components are updated to inform about ongoing handoff. Host base station passes the handover information to MS and handover is completed [2, 7, 10–12]. A handover can be hard handover, if physical switching of the frequency is involved; this kind of handover is also referred as break-before-make handoff. Another class is vertical handoff when handoff is performed between two different generations of communication networks like 2G to 3G /LTE [7–9].

The similarity in the handoff process encouraged us to investigate the process in isolation independently without considering the signaling being done to achieve transition in any generation of network. This simulation aims toward identifying the impact of other parameters on the performance of handover process. From metric point of view, call drop probability is having precedence over call blocking ratio; hence, in this experiment, call drop is considered to be the sole performance metric.

3 Simulation

We have simulated the process over the MASON, core framework in Java from George Mason University's Evolutionary Computation Laboratory. The simulation is set up using the parameters shown in Table 1.

The process is simulated in a space of 800×800 field using 2–4 base station. We have considered that base station controller is collocated with base station. Hence, the decision is coded with base station only. The channel is assumed to be noise-free, and there is no interference of any sort. Mobile nodes are aligned initially in small space which represents the road in urban cities. Mobile nodes move in straight line following water flow mobility model because in urban scenario, a MS cannot move haphazardly outside the bound of road network.

The handover is executed between threshold bounds defined as handover window. Beacon interval represents the time span between occurrences of resource facilitating the handover may be random access channel (RACH) or downlink dedicated control channel. Base stations are ideally located such that they overlap each other maximally to avoid shadow region between two cells and also provide complete coverage to their respective regions. Start factor defines the simulation step at which node start their execution. All the connected nodes are considered to be active nodes, and each base station is having a certain fixed number of channels to support handover.

The simulation has exhibited some interesting result which led to conventional as well as unusual inferences.

Table 1 Simulation setup parameter

Parameter	Value or range	Description
no_of_node	[2–500]	Number of nodes
no_of_bts	2/4	Total number of base stations used
staColors	Array of color	Color of the mobile nodes
ho_Channels	[3–15]	Handover channel as measure of percentage of total active call channels
beaconInterval	1	Interval of beaconing for the base station
Coverage	[100–800]	The coverage area of the base station. This the diameter of the space under the coverage of any base station
start_factor	5	The simulation start time factor used for each entity separately
ho_wnd_min	0.3 f	Minimum threshold for handover initiation
ho_wnd_max	2.0 f	Maximum threshold for handover termination
[XMIN–XMAX]	[0–800]	Bounds of X-axis
[YMIN–YMIN]	[0–800]	Bounds of Y-axis
Diameter	8	Size of mobile nodes and base stations
roadWidth	20	The width of road assumed
iniDensity	20	At initialization time nodes will be placed in space of (roadWidth × iniDensity)

3.1 Impact of redundant base station to lower the call drop

The redundancy of resources is often provisioned with an expectation toward ensuring consistent performance of the system. In case of cellular network, multiple base stations may be provisioned for covering the same region. The simulation is built with as many as four base stations completely covering the simulation area (Fig. 1).

Interestingly, the simulation results have inverted the entire hypothesis as there is a consistent call drop of about 50 % which was earlier absent when there were only two base stations to cover same geographical area (Fig. 3).

It can be seen from Fig. 3 that when there are the 100 users in network, i.e., 50 in either side, there are total 6 call drops contrarily; in Fig. 2, it is 110 call drops from total 200 users in network. It was found to against the general convention, the reason we find was something unusual and it is like that whenever a node moves from one end to another say along X-axis node, say base station A, finds perpendicularly located base station B on Y-axis closer than base station C on the other side along X-axis opposite to A. Therefore, node is transferred to the base station B as it is closer than C. When node further moves, it finds itself again in

Fig. 1 Cell layout with four base stations

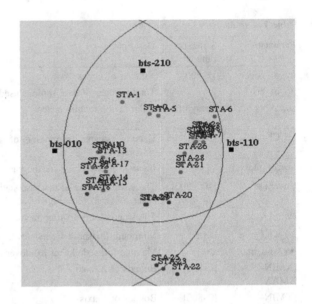

Fig. 2 Call drop with topology of four base stations

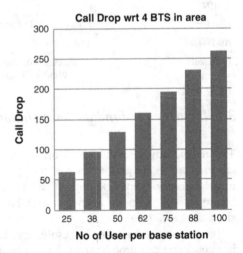

handover situation between B and C. However, by this time, the node has traveled through the handover window between B and C, leaving a smaller effective window for handover. It is quite obvious that smaller the window, higher will be the call drop (Fig. 4).

This problem has another perspective, with the advent of multiple generations of cellular network topologies like umbrella cell and co-located overlay. In both of these approaches, the overlying cells are entrusted with responsibility to manage load which underlying cell is incapable of handling. In case of higher rate mobility, maximum traffic is diverted to overlying cell. This process of load

Fig. 3 Call drop with topology of two base stations

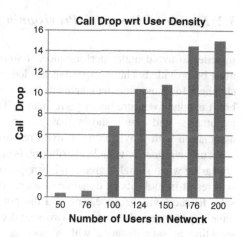

Fig. 4 Call drop with respect to threshold window with two base stations

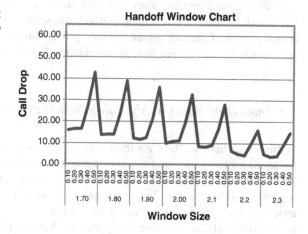

balancing is causing an avoidable congestion at overlying cell. Thus, even though service area is better covered, there is no guarantee for seamless connectivity.

It may be inferred from Fig. 4, that smaller the handover window size, more will be the call drops. In unban setups where connection density is often quite high, microcell configuration is most preferable as this allows for maximal frequency reuse. In microcell configuration, cells are at most 500 m–1 km in size. This leads to more frequent handover and larger possibility of call drops. Looking in Fig. 4, one may plan over a strategy to pre-initiate handover before it reaches the prescribed limits. Such strategies could be realized in reality because if, for example, the speed of a vehicle is larger than a limit, then it neither can stop instantaneously nor reverse the direction of its motion.

3.2 *Impact of Resource Provisioning on Throughput*

In order to avoid under performance, if some extra resource can be provisioned must be provided. This perspective has led to vast research in provision for extra channels for handover management. This can be done statically or dynamically, but it is done to ensure better performance. The simulation has totally rejected the hypothesis, and result shows that performance of the handover has no clear association with provisioning of extra resources.

In urban scenario, vehicle distribution is never Gaussian; rather, it is rectangular appearing when signal is green and disappearing for some time when it is red. This will result in mobility in dynamic cluster; group of indefinite size of people with safe distance will be appearing and disappearing together from the cell. As we have experienced also between two signal crossings, there is a bunch of people traveling at safe distance with almost same velocity. When bunch of people approaching handover boundary and node are more than the capacity provisioned then it exhibit an erratic relationship which can sole be defined with respect to the arrival pattern of the user. Since resource provisioning is done irrespective of this fact, hence, we cannot see any vibrant relationship (Fig. 5).

3.3 *Impact of Specific Signaling Schedules on Handover*

In this simulation, we have not implemented any sort of messaging system except for handover intimation and acknowledgment; therefore, we refer the signaling as beacon. Beacon in true sense is small signal carrying very minimum information sometime nothing at all. In our context, we have implemented beaconing as appearance of message from base station at regular interval measure with respect to simulation clock. To successfully execute handover two things are important firstly the cell identification broadcasts and other is appearance of RACH. Cell broadcast helps a mobile node know about the identification of cell and inform its parameter. RACH channel help in posting request to base station. RACH channel is usually accessed through contention process. The simulation results (Fig. 6) have appeared upside down with respect to conventional belief.

Fig. 5 Call drop with dedicated handover channels

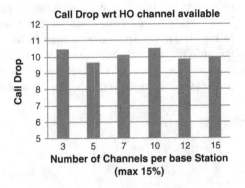

Fig. 6 Beaconing interval
and call drop

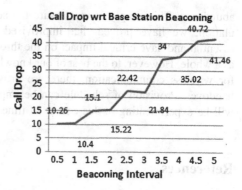

Call Drop wrt Base Station Beaconing

Let us understand this with a simple example. RACH is a logical channel and is defined in almost all wireless communication standards. RACH is appearing at periodical interval to facilitate communication between system and user mobile node, and same is used for handover messaging. If say this channel is appear at every T1 time and its one extra presence can be provisioned within T1 interval, then theoretically we can reduce the call drop probability to half. Similarly, if cell identification broadcasting appears in microcell in same way as it is schedule in macrocell, then there are chances that a node can pass by a cell without actually listening to the broadcast as a result call will be dropped. Therefore, we have to investigate properly to finally suggest a signaling schedule.

4 Conclusions

The experiments conducted here offer twofold findings. On one side, these throw light on association with not much explored attributes like beaconing interval. On the other side, one may infer from the experimental findings that factors like handover channel provisioning at base station are not the sole dependencies for better call admission. These hidden relationships posses a new kind of challenges and afresh investigation to tackle the situation.

Threshold window is the period between the threshold limits encapsulating the hysteresis in between. Existing convention is that larger the window, smaller will be the call drops and same is verified during the simulation. Similarly, the hypothesis on higher velocity, deprecation in handover priority, and higher user density results into higher call drop have been verified by the simulation experiments. The persistence of hypotheses during simulation has established the worth of our simulation setup and platform.

On the other hand, in these simulation experiments, we have proven experimentally that some of the existing conventions are not always augmenting the performance of the system rather hampering it drastically in cases. Experiment has shown that how intelligently network resources are made available and in abundance. This simply cannot assure the success of process. Through our simulation, we have proved that improvised signaling schedules are extremely essential and have critical impact on the throughput of the process and on system as a whole. However, to the best of our knowledge, in none of the existing standard for wireless communication, there are provisions for specialized signaling for handover. Provision of schedule can be implemented in several ways which we will be experimenting in due course of time and establish in the real terms.

References

1. Ahmad, M.H.: Call admission control in wireless networks: a comprehensive survey. IEEE Commun. **7**(1) (2005)
2. Tekinay, S., Jabbari, B.: Handover and channel assignment in mobile cellular networks. IEEE Commun. Mag. **29**, 42–46 (1991)
3. Ekiz, N., Salih, T., Küçüköner, S., Fidanboylu, K.: An overview of handoff techniques in cellular networks. Int. J. Inf. Technol. **2**(3) (2005)
4. Akhila, S, Kumar, S.: Handover in GSM networks. In: IEEE Proceedings of Fifth International Conference on MEMS NANO and Smart Systems, pp. 142–145 (2009)
5. Abdulova, V., Aybay, I.: Predicative mobile-oriented channel reservation schemes in wireless networks. Wirel. Netw. doi:10.1007/s11276-010-0270-2 (published online August 2010)
6. Lee, W., Cho, D.-H.: Group handover scheme using adjusted delay for multi-access networks. In: Proceedings of ICC 2010, IEEE, Online: 978-1-4244-6404-3
7. ETSI Technical Specification: Digital cellular telecommunications system (Phase 2+); universal mobile telecommunications system (UMTS); handover procedures, 3GPP TS 23.009 version 11.2.0 Release 11 (2013)
8. Guerin, R.A.: Queueing-blocking system with two arrival stream and guard channels. IEEE Trans. Commun. **36**, 153–163 (1988)
9. Zhang, Y., Salari, E.: A hybrid channel allocation algorithm with priority to handoff calls in mobile cellular networks. Comput. Commun. **32**(5), 880–887 (2009)
10. Madan, B.B., Dharmaraja, S., Trivedi, K.S.: Combined guard channel and mobile-assisted handoff for cellular networks. IEEE Trans. Veh. Technol. **57**(1) (2008)
11. ETSI Technical Specification: LTE; evolved universal terrestrial radio access (E-UTRA) and evolved universal terrestrial radio access network (E-UTRAN); overall description Stage 2, 3GPP TS 36.300 version 11.8.0 Release 11 (2014)
12. Ray, S.K., Pawlikowski, K., Sirisena, H.: Handover in mobile WiMAX networks: The state of art and research issues. IEEE Commun. Surv. Tutor. **12**(3), 376–399 (2010)

Author Index

© Springer India 2015
R. Chaki et al. (eds.), *Applied Computation and Security Systems*, Advances in Intelligent
Systems and Computing 304, DOI 10.1007/978-81-322-1985-9